Introduction

Genetics: PreTest® Self-Assessment and Review gives medical students, as well as physicians, a comprehensive and convenient instrument for self-assessment and review. The 500 questions parallel the format and degree of difficulty of the questions contained in Step 1 of the United States Medical Licensing Examination (USMLE) as well as the Foreign Medical Graduate Examination in the Medical Sciences (FMGEMS).

Each question in this book is accompanied by an answer, a paragraph explanation, and a specific page reference to a textbook. A bibliography listing the sources used in the book follows the last chapter.

Perhaps the most effective way to use this book is to allow yourself one minute to answer each question in a given chapter; as you proceed, indicate your answer beside each question. By following this suggestion, you will be approximating the time limits imposed by the board examinations previously mentioned.

When you finish answering the questions in a chapter, you should then spend as much time as you need verifying your answers and carefully reading the explanations. Although you should pay special attention to the explanations of questions you answered incorrectly, you should read *every* explanation. The authors of this work have designed the explanations to reinforce and supplement the information tested by the questions. If, after reading the explanations for a given chapter, you feel you need still more information about the material covered, you should consult and study the references indicated.

Genetics

PreTest®
Self-Assessment
and Review

Genetics

PreTest®
Self-Assessment
and Review

Edited by

Golder Wilson, M.D., Ph.D.
Professor of Pediatrics
The University of Texas Southwestern Medical School
Dallas, Texas

Janice Finkelstein, M.D.
Assistant Professor of Pediatrics
The Johns Hopkins University School of Medicine
Baltimore, Maryland

McGraw-Hill, Inc.
Health Professions Division/PreTest Series

New York   St. Louis   San Francisco   Auckland
Bogotá   Caracas   Lisbon   London   Madrid
Mexico   Milan   Montreal   New Delhi   Paris
San Juan   Singapore   Sydney   Tokyo   Toronto

Genetics: PreTest® Self-Assessment and Review
Copyright © 1993 by McGraw-Hill, Inc. All rights reserved. Printed in the
United States of America. Except as permitted under the Copyright Act of 1976,
no part of this publication may be reproduced or distributed in any form or by
any means, or stored in a data base or retrieval system, without the prior written
permission of the publisher.

1 2 3 4 5 6 7 8 9 0   DOCDOC   9 8 7 6 5 4 3

ISBN 0-07-052009-7

The editors were Gail Gavert and Bruce MacGregor.
The production supervisor was Gyl A. Favours.
This book was set in Times Roman by Compset, Inc.
R.R. Donnelley & Sons was printer and binder.

Library of Congress Cataloging-in-Publication Data

Genetics : PreTest self-assessment and review /
  Golder N. Wilson, Janice E. Finkelstein—1st ed.
      p.   cm.
  Includes bibliographical references.
  ISBN 0-07-052009-7
    1. Medical genetics—Examinations, questions, etc.
  2. Human genetics—Examinations, questions, etc.
  I. Wilson, Golder N.   II. Finkelstein, Janice E.
    [DNLM: 1. Genetics—examination questions.
    QH 431 G3293]
  RB155.G395   1993
  616'.042'076—dc20
  DNLM/DLC
  for Library of Congress   92-49120
                                                                CIP

Contents

Basic Genetics

DIRECTIONS: Each question below contains five suggested responses. Select the **one best** response to each question.

1. Mendel's laws apply to every statement below EXCEPT

(A) many traits are determined by a pair of hereditary units (*genes* or *alleles* in modern terminology)
(B) gametes (ova or sperm) each receive one of the paired alleles
(C) there is random sorting of alleles into ova and sperm
(D) alleles at loci on the same chromosome may segregate together
(E) the pair of alleles is reconstituted by zygote formation

2. A couple who both have Bb genotypes at a locus will produce zygotes in which of the following ratios?

(A) 1BB:1Bb:1bb
(B) 2BB:1Bb:1bb
(C) 1BB:2Bb:1bb
(D) 1BB:2Bb:2bb
(E) 1BB:3Bb

3. What proportion of families with three children will have all boys?

(A) 1/3
(B) 1/6
(C) 1/8
(D) 1/12
(E) 1/64

4. What is the baseline risk for congenital malformations in the average pregnancy?

(A) 2/10,000
(B) 2/1000
(C) 2/100
(D) 2/10
(E) 2/5

5. The average incidence of common single malformations such as cleft palate or spina bifida is

(A) 1/10,000
(B) 1/5000
(C) 1/1000
(D) 1/500
(E) 1/100

DIRECTIONS: Each group of questions below consists of lettered headings followed by a set of numbered items. For each numbered item select the **one** lettered heading with which it is **most** closely associated. Each lettered heading may be used **once, more than once, or not at all.**

Questions 6–10

Match each term with its partial definition.

(A) Heritable change in DNA
(B) Alternative form of a gene
(C) Position of gene on chromosome
(D) One gene, multiple effects
(E) Complete set of genes in cell or organism

6. Genome

7. Allele

8. Mutation

9. Locus

10. Pleiotropy

Questions 11–15

Match each term with the appropriate description.

(A) Cosegregation of alleles
(B) One phenotype, multiple genotypes
(C) Nonrandom allele association
(D) One locus, multiple mutant alleles
(E) One locus, multiple normal alleles

11. Polymorphism

12. Linkage

13. Linkage disequilibrium

14. Genetic heterogeneity

15. Allelic heterogeneity

Questions 16–20

Match each description with the correct disorder.

(A) Chromosomal disorders
(B) Single gene defects
(C) Multifactorial disorders
(D) Somatic cell genetic defects
(E) Mitochondrial disorders

16. Mendelian inheritance

17. Human cancers

18. Most common type of human genetic disease

19. Major cause of miscarriages

20. Maternal derivation

Questions 21–23

Match the number of possible alleles at a single locus with the number of genotypes that could result.

(A) Three possible genotypes
(B) Six possible genotypes
(C) Nine possible genotypes
(D) Ten possible genotypes
(E) Sixteen possible genotypes

21. Two alleles

22. Three alleles

23. Four alleles

Questions 24–28

For each of the categories of genetic disease, match the approximate incidence.

(A) 0.3 to 0.4 percent
(B) 0.5 to 0.7 percent
(C) 3 to 5 percent
(D) 10 to 25 percent
(E) 50 percent

24. Chromosomal disorders at birth

25. Chromosomal disorders in persons age 25 or younger

26. Mendelian (single gene) disorders in persons 25 or younger

27. Multifactorial (polygenic) disorders in persons 25 or younger

28. Genetic disorders as the cause for admission to a pediatric hospital

Questions 29–40

Match the following with the appropriate pedigree symbols indicated in the figure.

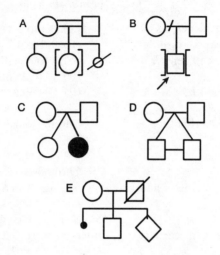

29. Consanguinity

30. Monozygotic twins

31. Offspring, sex unspecified

32. Dizygotic twins

33. Adopted into family

34. Adopted out of family

35. Deceased adult

36. Affected individual

37. Proband

38. Prenatal death

39. Miscarriage

40. Divorce

Basic Genetics
Answers

1. The answer is D. *(Gelehrter, pp 27–28. Thompson, 5/e. p 2.)* Mendel's laws assert that (1) traits reflect the inheritance of units called *genes*, (2) genes come in pairs that separate into different gametes at meiosis (segregation), and (3) gene pairs segregate independently of one another. Cosegregation of syntenic alleles (alleles on the same chromosome) violates Mendel's law of independent assortment, although it represents the phenomenon of genetic linkage. Mendel's pea loci were on different chromosomes and thus were inherited independently. More recently discovered inheritance mechanisms that violate Mendel's laws include genomic imprinting, mitochondrial inheritance, and germinal mosaicism.

2. The answer is C. *(Gelehrter, pp 27–30. Thompson, 5/e. p 2.)* Possible combinations of alleles in gametes form the basis of most genetic risk calculations and can be simplified by using the Punnett square.

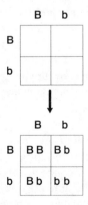

The parental alleles are aligned on two sides of the square and the potential combinations of gametes are listed in each compartment. Ratios of the genotypes can then be read directly.

3. The answer is C. *(Gelehrter, pp 30–31. Thompson, 5/e. pp 146–147.)* Each pregnancy may be taken as an independent event with a 1/2 chance of a male offspring. Therefore, the chance that three separate pregnancies will result in the birth of a male child is $1/2 \times 1/2 \times 1/2$, or 1/8.

4. The answer is C. *(Gelehrter, pp 27–30. Thompson, 5/e. p 2.)* The incidence of major congenital anomalies or genetic disease at birth is at least 2 to 3 percent of all pregnancies. If children are followed until age 7, the rate approaches 10 percent because of problems not appreciated earlier (e.g., hearing or heart defects). Genetic disorders affect every organ system and cause a significant portion of infant mortality.

5. The answer is C. *(Gelehrter, pp 27–30. Thompson, 5/e. p 2.)* Common birth defects have an incidence of about 1/1000 live births. Examples include neural tube defects (spina bifida/myelomeningocele), most congenital heart defects, isolated cleft palate and cleft lip with cleft palate, congenital dislocated hip, and hydrocephalus. Precise incidence figures vary by defect and population. While a 1/1000 incidence may not seem "common," this translates to 300 cases per year in a state such as Texas (300,000 annual births) and 3000 annual cases in the U.S. (3 million annual births).

6–10. The answers are 6-E, 7-B, 8-A, 9-C, 10-D. *(Thompson, 5/e. pp 427–442.)* Each organism is defined by a characteristic number and arrangement of genes that is termed the *genome*. Genes are the basic unit of hereditary material and occupy a specific address, or locus, on a chromosome. Mutations alter the sequence of DNA nucleotides that compose a gene; variant genes occupying the same chromosomal locus are termed *alleles*. Mutant genes that change several body structures or functions are called *pleiotropic*.

11–15. The answers are: 11-E, 12-A, 13-C, 14-B, 15-D. *(Gelehrter, pp 299–311.)* Polymorphic loci have multiple alleles because of DNA sequence variation. The DNA sequence changes may involve restriction sites (restriction fragment length polymorphisms, RFLPs), repeated segments (variable number of tandem repeats, VNTRs), or expressed regions (protein polymorphisms). Polymorphisms allow deduction of relationships between loci (i.e., cosegregation or linkage of particular alleles in a family) and between alleles (i.e., disequilibrium of allele with allele or allele with disease). Different mutant alleles may cause indistinguishable phenotypes (allelic heterogeneity) as may mutations at different loci (genetic heterogeneity).

16–20. The answers are: 16-B, 17-D, 18-C, 19-A, 20-E. *(Gelehrter, pp 3–5. Thompson, 5/e. pp 4–7.)* Genetic disorders may be classified in several major categories. Chromosomal disorders are caused by the deletion or duplication of either pieces of chromosomes or entire chromosomes and are a common cause of miscarriage. Single gene disorders, also known as *Mendelian disorders,* are due to defects in single genes. Multifactorial disorders, the most common type of human genetic disease, represent the composite effects of multiple genes, each of which contributes a minor component to the disorder. Environmental factors also play a role in multifactorial disorders. Many common diseases, such as coronary artery disease and diabetes mellitus, are multifactorial disorders. In human cancers, malignancy may result from a mutation in a gene involved in regulation of cell growth. These mutations occur in specific somatic cells rather than in all cells of the body. Defects in mitochondrial DNA may result in disease. Since mitochondria are cytoplasmic organelles inherited via the cytoplasm of the ovum, these disorders may be maternally inherited.

21–23. The answers are: 21-A, 22-B, 23-D. *(Gelehrter, pp 49–52. Thompson, 5/e. p 179.)* For a two-allele system (e.g., A and B), three genotypes are possible (AA, AB, BB). In a three-allele system (A, B, and C), six genotypes are possible (AA, AB, AC, BB, BC, and CC). For a four-allele system, there are ten possible genotypes (AA, AB, AC, AD, BB, BC, BD, CC, CD, and DD). In general, if a is the number of alleles, a is also the number of homozygotes and $a(a - 1)/2$ is the number of heterozygotes. The total number of genotypes is the sum of homozygotes and heterozygotes, or $a + a(a - 1)/2 = (a + a^2)/2$.

24–28. The answers are: 24-B, 25-A, 26-A, 27-C, 28-D. *(Gelehrter, pp 27–30. Thompson, 5/e. p 2.)* Chromosomal disorders, because they often produce severe phenotypes, have a larger incidence at birth (0.5 to 0.7 percent in newborn surveys) than in older age groups. Mendelian disorders will increase in incidence with age because they may produce subtle phenotypes in the newborn but have major consequences in later life (e.g., Marfan syndrome, familial hypercholesterolemia). Multifactorial disorders include common birth defects and diseases with genetic predisposition, such as schizophrenia, alcoholism, and diabetes mellitus. Current incidence figures are undoubtedly underestimates that will increase once widespread genetic screening is available for complex diseases.

29–40. The answers are: 29-A, 30-D, 31-E, 32-C, 33-A, 34-B, 35-E, 36-C, 37-B, 38-A, 39-E, 40-B. *(Gelehrter, pp 27–29. Thompson, 5/e. pp 53–57.)* A pedigree is a scientific genealogy in which individuals and relationships are symbolized precisely. Each row of symbols represents a generation, and the proband (or propositus), who is identified by an arrow, is the person first ascertained with the disease of interest or the person who requests information. The latter individual is also termed a *consultand.* Males are represented by squares and females by circles. A diamond may be used when sex is unknown. Symbols representing affected individuals are filled in. A horizontal line connecting two individuals represents a mating; a double horizontal line indicates consanguinity, or a mating of related individuals. Vertical lines denote offspring. Diagonal lines with a single point of origin denote twins, with monozygotic (or "identical") twins connected with a horizontal line. Death is indicated by a slash through the symbol. Brackets are used to represent adoption with the direction of the brackets showing whether the individual was adopted in or out.

Chromosomal Inheritance

DIRECTIONS: Each question below contains five suggested responses. Select the **one best** response to each question.

41. Clinical indications for karyotyping include all the following EXCEPT

(A) multiple malformations in a newborn
(B) single malformation in a newborn
(C) mental retardation of unknown etiology
(D) offspring with chromosomal rearrangement
(E) recurrent pregnancy loss

42. Chromosomal imbalance is most frequent during which of the following stages of human development?

(A) Embryonic
(B) Fetal
(C) Neonatal
(D) Childhood
(E) Adult

43. The cell labeled A in the figure below is best described by the terms

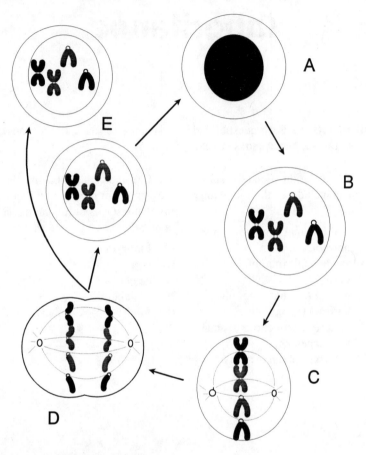

(A) meiotic I interphase, haploid
(B) mitotic telophase, diploid
(C) meiotic I metaphase, diploid
(D) mitotic interphase, diploid
(E) mitotic anaphase, diploid

DIRECTIONS: Each group of questions below consists of lettered headings followed by a set of numbered items. For each numbered item select the **one** lettered heading with which it is **most** closely associated. Each lettered heading may be used **once, more than once, or not at all.**

Questions 44–47

Match each clinical situation below with the appropriate risk figure.

(A) 1/10,000
(B) 1/600
(C) 1/100
(D) 1/10
(E) 1

44. The risk for a newborn to have Down syndrome

45. The risk for a balanced translocation carrier to have a child with unbalanced chromosomes

46. The theoretical risk for a 21/21 translocation carrier to have a child with Down syndrome

47. The risk for parents of a trisomy 21 child to have a second offspring with a chromosomal abnormality

Questions 48–51

For each description, select the appropriate karyotype.

(A) 46,XY
(B) 23,X
(C) 69,XXY
(D) 47,XX+21
(E) 92,XXXX

48. Haploid

49. Diploid

50. Aneuploid

51. Triploid

Questions 52–56

Indicate the physical (chromosomal) basis for each of the genetic observations below.

(A) Meiotic recombination
(B) Expansion of repetitive DNA families
(C) Separation of chromosomal homologues (reduction division)
(D) 47,XXY karyotype
(E) 47,XYY karyotype

52. Mendel's law of independent assortment

53. Sister chromatid exchange

54. Genetic distance

55. Nondisjunction at meiosis I or II

56. Nondisjunction at meiosis II only

Questions 57–61

Match the conditions below with their cytogenetic notation represented by letters A through E.

(A) 47,XX,+21
(B) 45,X
(C) 46,XX,r(15)
(D) 47,XY,+21
(E) 45,XX-21

57. Male with trisomy 21 (Down syndrome)

58. Patient with ring 15

59. Female with trisomy 21

60. Female with monosomy X (Turner syndrome)

61. Patient with monosomy 21

Questions 62–69

Match the terms below with the chromosome structures represented by the letters in the figure.

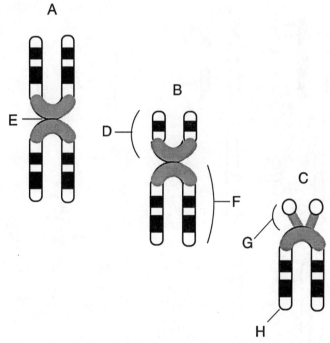

62. Submetacentric chromosome

63. Short arm, or p

64. Acrocentric chromosome

65. Satellite

66. Centromere

67. Metacentric chromosome

68. Long arm, or q

69. Chromatid

Questions 70–74

Match the cytogenetic nomenclature with the lettered diagrams in the figure.

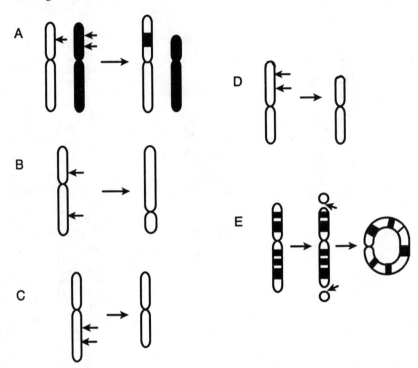

70. Ring chromosome

71. Interstitial deletion of short arm

72. Pericentric inversion

73. Insertional translocation (nonreciprocal)

74. Interstitial deletion of long arm

Questions 75–81

Match the following cytogenetic notations with the lettered diagrams in the previous question group.

75. 46,XX,-7,-9, + der(7)(7pter→7p13::9p25→9p13: :7p13→7qter), + del(9)(p25p13)

76. 46,XX,r(15)(p25q23)

77. 46,XY,del(8)(p23p13)

78. 46,XY,inv(9)(p13q24)

79. 46,XX,del(6)(q15q25)

80. 46,XY,r(7)(p24q25)

81. 46,XX,inv(7)(p14q25)

Questions 82–86

Match the cytogenetic nomenclature with the lettered diagrams.

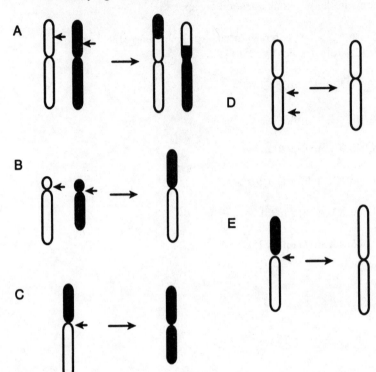

82. Isochromosome (short arm)

83. Paracentric inversion

84. Reciprocal translocation

85. Isochromosome (long arm)

86. Robertsonian translocation

Questions 87–94

Match the following cytogenetic notations with the diagrams in the previous question group. Assume that the abnormal homologue illustrated in the diagrams is accompanied by a normal homologue to generate the karyotype.

87. 46,XX,i(6p)

88. 46,X,i(Xp)

89. 46,XY,t(6;7)(p22;p15)

90. 45,XY,t(13p14p)

91. 46,XY,inv(8)(q14q24)

92. 46,XY,i(9q)

93. 46,XY,-3,-5,+der(3), +der(5)t(2,5)(p23;p25)

94. 45,XX,t(14p15p)

Questions 95–101

Match each cytogenetic notation with the appropriate phenotype.

(A) Down syndrome
(B) Translocation carrier with normal phenotype
(C) Patau syndrome (the phenotype produced by an extra copy of chromosome 13)
(D) Turner syndrome
(E) Down syndrome with possible atypical phenotype

95. 45,XX,t(21q21q)

96. 46,XX,t(21q21q)

97. 47,XY,+21

98. 47,XX,+der(2)(2pter→2q11::21p11→21qter)

99. 46,XY,t(p13p13)

100. 46,X,i(Xq)

101. 45,X

Questions 102–109

Match each description below with the correct stage of meiosis depicted in the figure.

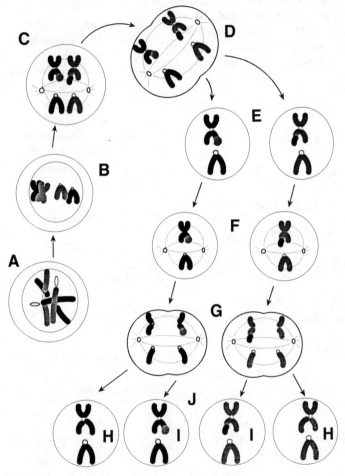

102. Tetrads

103. Zygotene and synapsis

104. Spermatogonia

105. Oocytes

106. Meiosis II metaphase

107. Segregation of homologues (bivalents)

108. Separation of chromatids

109. Recombinant gamete

Questions 110–118

Match each clinical diagnosis or cytogenetic notation with the appropriate karyotype. Answer E if none of the illustrations apply.

110. Male with Down syndrome

111. 69,XXY

112. Female with Down syndrome

113. Turner syndrome

114. 92,XXXY

115. Triploidy

116. Translocation Down syndrome

117. 45,X

118. Parental karyotypes should be obtained

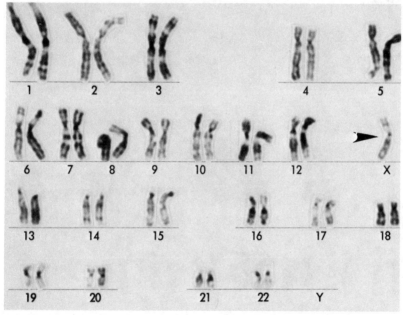

A

(Illustrations continue next page)

B

C

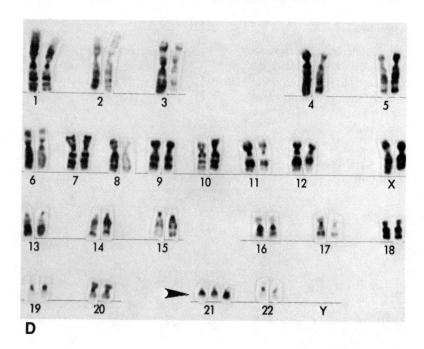

D

Questions 119–124

Match each description or cytogenetic notation with the appropriate karyotype. Answer E if none of the illustrations apply.

119. 18p − syndrome

120. 4p − , or Wolf-Hirschhorn, syndrome

121. Klinefelter syndrome

122. 47,XXY

123. 47,XYY

124. 13p −

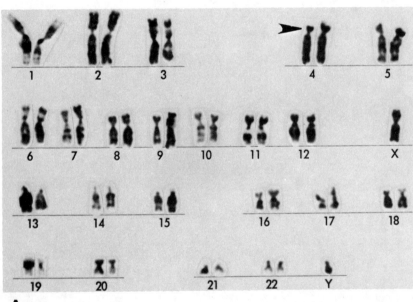

A

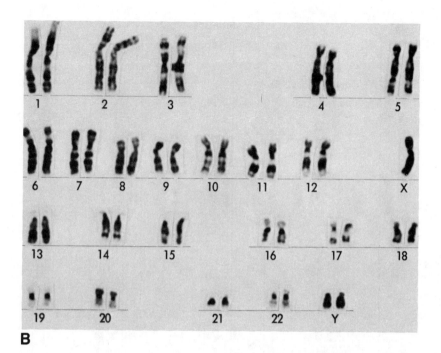

B

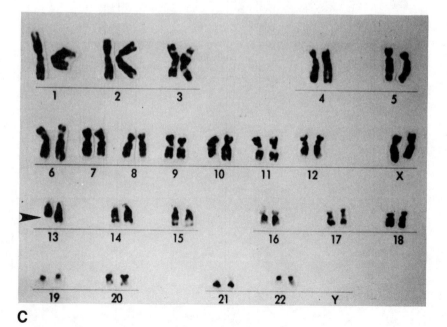

C

(Illustrations continue next page)

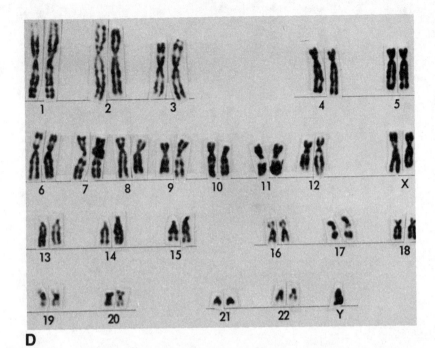

D

Questions 125–128

Match each description or cytogenetic notation with the correct illustration. Answer E if none of the illustrations apply.

125. 18q– syndrome

126. Cri du chat syndrome

127. Balanced translocation carrier

128. Translocation Down syndrome

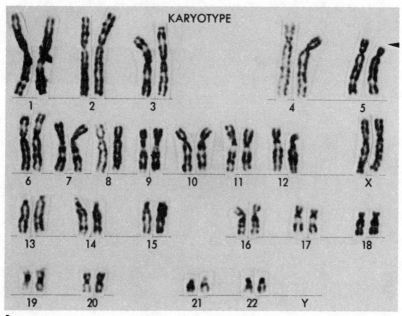

A

(Illustrations continue next page)

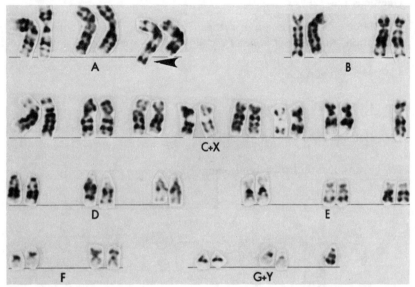

B

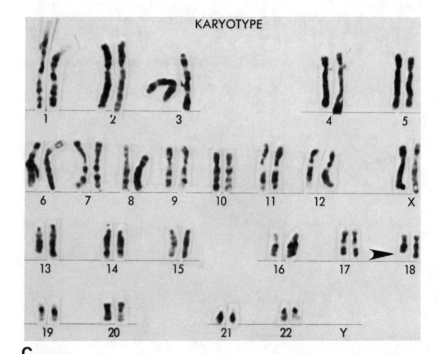

KARYOTYPE

C

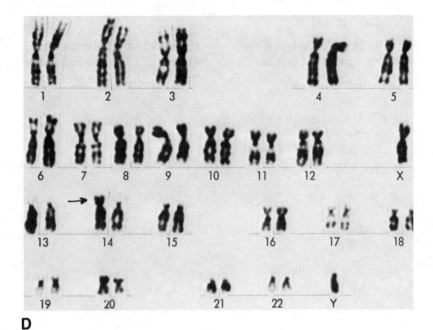

D

Questions 129–132

Match each numbered item with the correct description.

(A) Partial trisomy
(B) Satellite
(C) High probability of abnormal phenotype
(D) Heteromorphism
(E) None of above

129. 9qh+

130. 16qh+

131. 21p+

132. 22s+

Questions 133–137

Match each numbered item with the correct lettered description.

(A) One Barr body
(B) No Barr bodies
(C) Four inactive X chromosomes
(D) Two inactive X chromosomes
(E) None of above

133. Turner syndrome

134. 48,XXXX

135. 48,XXXY

136. 49,XXXXX

137. 46,XX

Chromosomal Inheritance

Answers

41. The answer is B. *(Gelehrter, p 187.)* The hallmarks of chromosomal disease in children are multiple malformations and mental retardation. Frequently growth is impaired and the facial appearance is unusual. Parents of children with chromosomal rearrangements (translocations, deletions, duplications) must be karyotyped to determine if they are carriers of balanced translocations. Approximately 5 percent of couples with more than three first-trimester abortions will have balanced chromosomal rearrangements as a cause for pregnancy loss or infertility. Chromosomal analysis should also be considered in patients with unexplained mental retardation, particularly if they have an unusual appearance or a family history of mental retardation. The child with a single birth defect (e.g., cleft palate, dislocated hip) will rarely have a chromosomal abnormality unless there are undetected minor or internal anomalies that compose a malformation syndrome.

42. The answer is A. *(Thompson, 5/e. pp 215–216.)* Chromosomal aberrations occur in approximately 1 in 200 liveborn infants, 4 percent of stillbirths, and 50 percent of all first-trimester abortuses. At least 15 to 20 percent of conceptions result in spontaneous abortion (miscarriage) during the first trimester (<12 weeks' gestation). Although the exact frequency of chromosomal anomalies in human embryos (<8 weeks' gestation) is unknown, the numbers above indicate a substantial frequency of at least 7.5 percent.

43. The answer is D. *(Gelehrter, pp 20–21.)* The figure depicts mitosis with its stages of interphase (A), prophase (B), metaphase (C), telophase (D), and division to yield daughter cells (E). Mitosis is the mechanism by which somatic cells duplicate each pair of chromosomes so as to produce identical copies in diploid daughter cells. At mitotic interphase, the diploid chromatin is dispersed and actively transcribed prior to DNA replication and condensation.

44–47. The answers are: 44-B, 45-D, 46-E, 47-C. *(Gelehrter, chap 8. Thompson, 5/e. chap 9.)* The incidence of Down syndrome at birth is approximately 1 in 600 liveborn children, with 95 percent trisomy 21. About 4 percent of patients with Down syndrome have translocations that mandate parental karyotyping to determine if one of the parents is a balanced translocation carrier. The remaining 1 percent are mosaics, which means that certain tissues are mixtures of trisomy 21 and normal cells. Translocation carriers have a 5 to 20 percent risk for unbalanced offspring with female carriers in general at higher risk than male carriers. Offspring of translocation 21/21 carriers should in theory all have Down syndrome, although in practice some carriers have had normal children. The empiric risk for parents with a trisomy 21 child is 1/100 for a second child with chromosomal aneuploidy.

48–51. The answers are: 48-B, 49-A, 50-D, 51-C. *(Gelehrter, pp 163–164.)* There are 23 different chromosomes in normal human gametes: 22 autosomes and 1 X or Y sex chromosome. At fertilization, normal haploid gametes unite to form a diploid zygote with 46 chromosomes (euploidy). Errors in meiosis may produce gametes with one missing or extra chromosome, which after fertilization generate an embryo with 45 or 47 chromosomes (aneuploidy). Most aneuploid embryos abort in the first trimester, but selected trisomies and monosomies survive. Rarely, gametes contain an additional chromosome complement such that they have 2n or 3n chromosomes where n is 23. The resulting embryo will have 3n (69) or 4n (92) chromosomes and produce a triploid or tetraploid fetus with severe birth defects. Cytogenetic notation always lists the chromosome number first, followed by a comma and the sex chromosomes. A haploid gamete is represented as 23,X or 23,Y; a normal person as 46,XX or 46,XY; and a triploid person as 69,XXX, 69,XXY, or 69,XYY. Examples of aneuploid persons include 47,XX+21 (female with Down syndrome) and 45,X (Turner syndrome).

52–56. The answers are: 52-C, 53-B, 54-A, 55-D, 56-E. *(Thompson, 5/e. pp 22–30.)* Mendel's laws of segregation and independent assortment received direct physical verification when technology allowed visualization of chromosomal segregation during meiosis. Meiosis is most easily understood by considering the sex chromosomes during spermatogenesis. Segregation of X and Y homologues occurs at meiosis I followed by segregation of sister chromatids at meiosis II. Failure of segregation (nondisjunction) at meiosis I will generate only abnormal XY gametes, while nondisjunction at meiosis II can generate either XY, XX, or YY gametes. Thus, 47,XXY can result from nondisjunction at meiosis I or II, while 47,XYY can only result from nondisjunction at meiosis II.

Crossing-over (recombination) between chromosomal homologues prior to meiosis I will exchange parental alleles and allow calculation of distances between loci based on recombination frequency. Crossing-over (exchange) between sister chromatids during meiosis II or mitosis can rearrange alleles on one chromosomal homologue and lead to expansion of DNA repeats by misalignment.

57–61. The answers are: 57-D, 58-C, 59-A, 60-B, 61-E. *(Gelehrter, p 164.)* Cytogenetic notation provides the chromosome number (e.g., 46), the sex chromosomes, and a shorthand description of anomalies. An example would be 47,XX, + 13 to indicate a female with trisomy 13. Chromosome numbers of 45 and 47 will thus indicate aneuploidy with missing and extra chromosomal material, respectively. Ring chromosomes (r) occur when a crossover joins the tips of the long and short arms.

62–69. The answers are: 62-B, 63-D, 64-C, 65-G, 66-E, 67-A, 68-F, 69-H. *(Gelehrter, p 164.)* Chromsomes are routinely studied at mitotic metaphase when they consist of two sister chromatids joined at a centromere. The drawings in the question are idiograms that represent the banded metaphase chromosomes after hypotonic treatment, spreading on slides, and photomicroscopy. The short, or "petite," arm is positioned above the centromere by convention and is represented by a "p." The long arm is called "q" merely because that is the next letter in the alphabet. Chromosomes are metacentric, submetacentric, or acrocentric according to whether the centromere is at the center, off center, or at the very tip (*acro* means *end*) of the chromosome, respectively. In humans, the acrocentric chromosomes contain repetitive satellite and ribosomal DNA at their tips, which forms satellites (G).

70–74. The answers are: 70-E, 71-D, 72-B, 73-A, 74-C. *(Gelehrter, p 164.)* Ring chromosomes, interstitial deletions, and pericentric inversions are examples of intrachromosomal rearrangements in which breaks or crossovers unite different regions of the same chromosome. The inverted segment contains the centromere in *peri*centric inversions but does not in *para*centric inversions. Robertsonian translocations join acrocentric chromosomes by breakage and reunion of their short arms.

75–81. The answers are: 75-A, 76-E, 77-D, 78-B, 79-C, 80-E, 81-B. *(Gelehrter, pp 163–171. Thompson, 5/e. pp 201–214.)* The abbreviations *inv, dup, del,* and *r* describe inverted, duplicated, deleted, and ring chromosome segments, respectively. The segment involved is delimited by the bands at its borders. Bands are numbered according to their distance from the centromere: band 3p3 will be distal (nearer the telomere) and band 3p1 proximal (nearer the centromere) on the short arm of chromosome 3. Better techniques reveal subbands within larger bands that are numbered similarly (e.g., 3p11, 3p12). Shorthand notation is adequate for simple rearrangements as with ring 7 in question 80. The notation indicates that chromosome 7 has broken at bands p24 and q25 and rejoined as a ring. More complex rearrangements require longhand nomenclature exemplified by the insertional translocation described in question 75. Each segment of the derived (der) chromosome is symbolized including short-arm telomere (pter), long-arm telomere (qter), breakpoints (::), and intact regions ($\rightarrow$).

82–86.   The answers are: 82-C, 83-D, 84-A, 85-E, 86-B. *(Gelehrter, p 164.)* Reciprocal translocations (A) involve exchange of segments between two chromosomes, while Robertsonian translocations (B) involve joining of two acrocentric chromosomes by breakage and reunion of their short arms. Translocations producing no duplication/deficiency of chromosomal material are called *balanced,* and carriers of balanced translocations will have normal phenotypes unless the translocation alters expression of an important gene at the breakpoint region. Carriers of balanced reciprocal translocations have a normal chromosome number (46), but carriers of balanced Robertsonian translocations will have 45 chromosomes. Isochromosomes involve duplication of the short (C) or long (E) arms, which produces perfectly metacentric chromosomes deficient in long- or short-arm material, respectively. The chromosomal segment that is reversed in paracentric inversions does not contain a centromere and thus produces a change in banding pattern but no change in shape (D).

87–94.   The answers are: 87-C, 88-C, 89-A, 90-B, 91-D, 92-E, 93-A, 94-B. *(Gelehrter, pp 163–171. Thompson, 5/e. pp 201–214.)* The abbreviations *i* and *t* describe isochromosomes and translocation chromosomes, respectively. Isochromosomes may duplicate the short arm (e.g., questions 87 and 88) or the long arm (e.g., question 92). These karyotypes are abnormal since long- or short-arm segments are missing. Reciprocal translocations involve exchange of segments between two chromosomes. A semicolon (;) indicates this exchange and is placed between the breakpoints (questions 89 and 93). Robertsonian translocations join two acrocentric chromosomes as in questions 90 and 94.

Note that this joining produces a chromosome number of 45 in a person with a normal karyotype. Reciprocal or Robertsonian translocations that result in no missing or extra chromosomal material are called *balanced*, and persons with balanced translocations are often called *translocation carriers*.

95–101. The answers are: 95-B, 96-A, 97-A, 98-E, 99-C, 100-D, 101-D. *(Gelehrter, pp 171–189. Thompson, 5/e. pp 201–214.)* The 21/21 translocation in question 95 is not associated with an additional normal 21 chromosome, as is indicated by the chromosome number of 45. In question 96, there is a normal 21 chromosome in addition to the 21/21 translocation chromosome; this person thus has Down syndrome rather than being a balanced translocation carrier with a risk for chromosomally abnormal offspring. The person in question 98 has an extra derived chromosome formed by translocation; the extra region includes the 21q21 segment that produces Down syndrome in addition to a duplicated region of chromosome 2 that has been implicated in a different pattern of birth defects. Deletion of the entire X chromosome or its short arm gives rise to the phenotype of Turner syndrome (questions 100 and 101). Deletion of the X long arm—46,X,i(Xp)—produces a milder phenotype.

102–109. The answers are: 102-C, 103-C, 104-A, 105-J, 106-F, 107-D, 108-G, 109-I. *(Gelehrter, pp 20–21.)* Diploid oogonia and spermatogonia (stage A) undergo two meiotic divisions that separate chromosomal homologues into haploid oocytes and sperm (stage J). At prophase of meiosis I, 4n copies of each chromosome are aligned in synapsis at the tetrad stage (C). The first meiotic division separates homologues (no division of centromeres), while the second distributes sister chromatids into haploid gametes (G). Chiasmata (visible bridges between chromosomes) form at the tetrad stage and produce an average 1 to 2 crossovers per chromosome to generate recombinant gametes (I).

110–118. The answers are: 110-C, 111-B, 112-D, 113-A, 114-E, 115-B, 116-C, 117-A, 118-C. *(Gelehrter, pp 159–189. Thompson, 5/e. pp 201–230.)* Interpretation of karyotypes begins with noting the arrangement of chromosomes singly (haploid as in 23,X gametes), in pairs (diploid as in 46,XX somatic cells), in threes (triploid as in 69,XYY hydatidiform moles), and in fours (tetraploid as in certain liver cells). The sex chromosomes are then noted as in 45,X (figure A), and each pair of homologues is inspected to recognize differences in shape or banding pattern.

The extra material on chromosome 14 in figure C might require clinical correlation or special techniques to demonstrate that it is derived from chromosome 21. Rearrangements in offspring require karyotyping of the parents to determine recurrence risk.

119–124. The answers are: 119-E, 120-A, 121-D, 122-D, 123-B, 124-E. *(Gelehrter, pp 159–189. Thompson, 5/e. pp 201–230.)* The shorthand nomenclature 18p − refers to deletion of a segment of undetermined length from the terminus of chromosome 18. The deletion illustrated in figure C is not of 18 and affects the long arm, not short arm, of chromosome 13. Children with Wolf-Hirschhorn syndrome have severe growth and mental retardation, a prominent nose and forehead that resembles a Greek warrior helmet, and multiple congenital anomalies. Klinefelter syndrome (47,XXY) involves tall stature, gynecomastia, small testes, and a eunuchoid habitus; these patients often exhibit abnormal behavior with school problems and poor motivation. Similar behavioral manifestations without a distinctive phenotype occur in the "supermale" 47,XYY syndrome.

125–128. The answers are: 125-C, 126-A, 127-E, 128-D. *(Gelehrter, pp 159–189. Thompson, 5/e. pp 201–230.)* Infants with cri du chat syndrome may present with a high-pitched "cat's cry," microcephaly, and several minor anomalies. They have a deletion of the short arm of chromosome 5 (5p −). Del(18q) patients also have growth failure and mild mental retardation. They are remarkable for speech delay. The translocation (14/21) chromosome in figure D is accompanied by two normal 21 chromosomes, which results in the three doses of chromosome 21 that will produce Down syndrome.

129–132. The answers are: 129-D, 130-D, 131-C, 132-B. *(Gelehrter, p 164. Thompson, 5/e. pp 201–204.)* Repetitive DNA is clustered at the centromeric regions of all chromosomes and can undergo amplification to produce heteromorphisms (literally "different shapes"). These heteromorphisms are very common at the centromeric regions of chromosomes 1, 9, and 16 and occasionally are amplified as very large segments (1qh +, 9qh +, and 16qh +) that can be distinguished from trisomic material by centromeric banding. The telomeres of acrocentric chromosomes contain repetitive and ribosomal DNA, which may amplify to form visible satellites (e.g., 13s +, 22s +). Although the notation *21p +* could refer to satellite amplification, it implies that the origin of extra material on the 21 short arm is not well defined. If this extra segment contains single-copy DNA (i.e., functional genes) from any autosome, then the patient is likely to have mental retardation with multiple major or minor anomalies.

133–137. The answers are: 133-B, 134-E, 135-D, 136-C, 137-A. *(Gelehrter, pp 176–187. Thompson, 5/e. pp 231–246.)* Barr bodies are the cytologic correlates of inactive X chromosomes. Dosage compensation in humans requires inactivation of all but one X chromosome in each cell so that males and females will have appropriate amounts of X chromosome gene products. Barr bodies or inactive X chromosomes will thus be one less than the total number of X chromosomes. Although the Barr body is visible as a dot near the inner side of the nucleus in many somatic cells, tests to determine the number of Barr bodies (e.g., buccal smears) are no longer considered useful or reliable.

Mendelian Inheritance

DIRECTIONS: Each question below contains five suggested responses. Select the **one best** response to each question.

138. All the following statements are true regarding autosomal dominant conditions EXCEPT

(A) they tend to have a vertical pattern in pedigree
(B) males and females are equally affected
(C) the less the reproductive fitness, the less likely it is that the case resulted from a new mutation
(D) they are often clinically variable
(E) they are pleiotropic

139. Autosomal recessive conditions are correctly characterized by which of the following statements?

(A) They are often associated with deficient enzyme activity
(B) Both alleles contain the same mutation
(C) They are more variable than autosomal dominant conditions
(D) Most persons do not carry any abnormal recessive genes
(E) Affected persons are likely to have affected offspring

140. A man is identified as having an X-linked dominant disorder. His daughter appears unaffected. Possible explanations include all the following EXCEPT

(A) Turner karyotype
(B) Nonpaternity
(C) lyonization
(D) inheritance of unaffected paternal allele
(E) back mutation

DIRECTIONS: Each group of questions below consists of lettered headings followed by a set of numbered items. For each numbered item select the **one** lettered heading with which it is **most** closely associated. Each lettered heading may be used **once, more than once, or not at all.**

Questions 141–148

Match the characteristics to the mode of inheritance.

(A) Autosomal dominant
(B) Autosomal recessive
(C) X-linked recessive
(D) Chromosomal
(E) Polygenic

141. When male-to-male transmission is observed, this mode is unlikely

142. Elevated maternal age is characteristic

143. Parents with three affected children have a higher recurrence risk than parents with two affected children

144. Elevated paternal age is characteristic

145. Variable expressivity is associated

146. Daughters of affected fathers are obligate carriers

147. Inborn errors of metabolism are associated

148. Consanguinity is characteristic

Questions 149–153

Match each numbered statement regarding X-linked inheritance with the correct lettered statement.

(A) The disorder is genetically lethal in males
(B) The condition nearly always manifests itself in heterozygous females
(C) The condition is lethal in utero in hemizygous males
(D) Male-to-male transmission is possible
(E) Unaffected males transmit the phenotype

149. Approximately two-thirds of mothers of affected males are carriers

150. Females are affected twice as often as males

151. Males are affected much more frequently than females

152. Affected females show increased frequency of spontaneous abortions

153. Affected females have fewer sons than daughters

Questions 154–158

Match the terms below with the appropriate description.

(A) 1 locus, 2 identical alleles
(B) 1 locus, 2 different mutant alleles
(C) 1 locus, 1 normal allele, 1 mutant allele
(D) 1 locus, 1 allele
(E) 2 loci, 4 different alleles

154. Homozygote

155. Heterozygote (carrier)

156. Compound heterozygote

157. Double heterozygote

158. Hemizygote

Questions 159–166

Match the members of a family with their degree of relatedness to the index case.

(A) First-degree relative
(B) Second-degree relative
(C) Third-degree relative
(D) None of the above

159. Brother

160. Son

161. Wife

162. Half-brother

163. First cousin

164. Maternal aunt

165. Paternal aunt

166. Father

Questions 167–170

The major blood group locus in humans produces types A (genotypes AA or AO), B (genotypes BB or BO), AB (genotype AB), or O (genotype OO). For each pair of parents, match the possible offspring.

(A) Type AB child
(B) Type B child
(C) Type O child
(D) None of above

167. Type A father, type O mother

168. Type O father, type AB mother

169. Type AB father, type O mother

170. Type A father, type A mother

Questions 171–173

Phenylketonuria (PKU) is an autosomal recessive disease that causes severe mental retardation if it is undetected. Two normal parents are told by their state neonatal screening program that their third child has PKU. Assume the initial screening is accurate and answer the questions below.

(A) 100 percent
(B) 67 percent
(C) 50 percent
(D) 25 percent
(E) Virtually 0

171. What is the risk for their next child to have PKU?

172. What is the risk of their oldest child to be a carrier for PKU?

173. What is the risk for the mother to be a carrier for PKU?

Questions 174–178

A couple presents for genetic counseling after their first child was born with the dwarfing syndome achondroplasia. You obtain the following family history: The husband (George) is the first-born of four male children, and George's next oldest brother has cystic fibrosis. The wife is an only child. The couple and both sets of grandparents have had no medical problems.

You should now draw the pedigree with the female member of any couple toward the left. Number generations with roman numerals and individuals with arabic numerals and indicate those affected with achondroplasia or cystic fibrosis. (Recall that achondroplasia is autosomal dominant and cystic fibrosis is autosomal recessive.) Match the individuals below with the appropriate description by referring to your numbered pedigree.

(A) Affected with cystic fibrosis
(B) Normal female
(C) Normal male
(D) Proband
(E) Two-thirds chance of being a carrier of cystic fibrosis

174. Individual III-1

175. Individual I-1

176. Individual II-1

177. Individual II-5

178. Individual II-3

Questions 179–183

Pedigrees I through III in the figure below represent families with retinitis pigmentosa (RP), a genetically heterogeneous eye disease that causes progressive visual impairment. Match the risks.

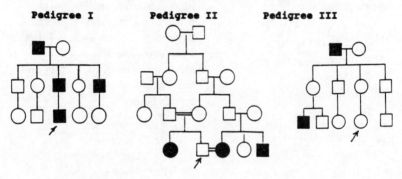

(A) 50 percent (1/2)
(B) 33 percent (1/3)
(C) 25 percent (1/4)
(D) 11 percent (1/9)
(E) Virtually 0

179. For I, risk for proband's son to have RP

180. For I, risk for proband's daughter to have RP

181. For II, risk for proband's child to have RP

182. For II, risk for proband's parents to have another RP child

183. For II, risk for proband to have affected child with RP if he had married his wife's unaffected sister

Questions 184–187

Refer again to the figure in the previous question group. Match the following risks regarding pedigree III.

(A) 50 percent (1/2)
(B) 33 percent (1/3)
(C) 25 percent (1/4)
(D) 11 percent (1/9)
(E) Virtually 0

184. Risk for proband's son to have RP

185. Risk for proband's daughter to have RP

186. Risk for affected males to have a child with RP

187. Risk for affected male's daughter to have a child with RP

Questions 188–190

Achondroplasia is an autosomal dominant condition that is characterized by dwarfism, prominent forehead, shallow nasal bridge, and short limbs. Because of support groups such as Little People of America, achondroplasts often meet each other and have families. Match the probabilities below.

(A) 100 percent
(B) 75 percent
(C) 50 percent
(D) 25 percent
(E) Virtually 0

188. The probability for the first child of achondroplastic parents to be affected (heterozygous or homozygous)

189. Following the birth of an affected child, the probability for the second child of achondroplastic parents to be affected

190. The probability for the first child to be affected if an achondroplast marries a normal person

Questions 191–195

Match the descriptions below with the appropriate risk figure.

(A) 100 percent
(B) 67 percent
(C) 50 percent
(D) 25 percent
(E) Virtually 0

191. The risk for a woman whose father has X-linked hemophilia to have an affected child

192. The risk for that same woman to have an affected son

193. The risk for a man with X-linked hemophilia to have an affected daughter

194. The risk for that same man to have a carrier daughter

195. The risk for that same man to have an affected son

Questions 196–199

Match the descriptions below with the appropriate term.

(A) Genetic heterogeneity
(B) Variable expressivity
(C) Germinal mosaicism
(D) Nonrandom lyonization
(E) Incomplete penetrance

196. A female carrier of hemophilia (X-linked disorder) has severe bleeding after routine cuts or abrasions

197. A grandson and paternal grandfather have ectrodactyly (autosomal dominant disorder with absent middle fingers), but the father has normal hands

198. An albino couple has a normal child (albinism is an autosomal recessive disorder)

199. A 90-year-old man with autosomal dominant neurofibromatosis has a son and grandson who died in their twenties from neural tumors

Questions 200–204

Tay-Sachs is an autosomal recessive disease that causes cherry red spots in the eye, "startle" responses in infancy, neurodegeneration, and death. Heterozygotes with an abnormal Tay-Sachs allele are termed *carriers*. Match the following individuals with their risks to be carriers.

(A) 100 percent
(B) 67 percent
(C) 50 percent
(D) 25 percent
(E) Virtually 0

200. Mother of an affected child

201. Sister of an affected child

202. First cousin of an affected child

203. Grandmother of an affected child

204. Half-brother of an affected child

Questions 205–209

Two normal parents have a child with sensorineural deafness. Many genetic forms of deafness exist, including autosomal dominant, recessive, and X-linked forms. Match the following assumptions with the risk for a second affected child.

(A) 50 percent
(B) 50 percent or virtually 0
(C) 25 percent
(D) 25 percent or virtually 0
(E) Virtually 0

205. Autosomal recessive inheritance

206. X-linked recessive inheritance

207. Autosomal dominant inheritance with complete penetrance

208. Autosomal dominant inheritance with incomplete penetrance

209. X-linked dominant inheritance

Questions 210–214

The pedigree shown in the figure below contains individuals with Charcot-Marie-Tooth (CMT) disease, a neurologic disorder that produces dysfunction of the distal extremities with characteristic footdrop. Match the individuals in the pedigree with their probability of having an affected child with CMT.

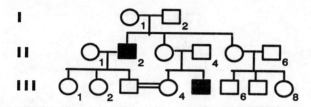

(A) 1 in 2
(B) 1 in 4
(C) 1 in 8
(D) 1 in 16
(E) Virtually 0

210. Individual II-2

211. Individual II-3

212. Individual II-5

213. Individual III-4

214. Individual III-8

Questions 215–219

The pedigree shown in the figure below also contains individuals with Charcot-Marie-Tooth (CMT) disease. However, this variant of CMT only becomes manifest in the late twenties. Match the individuals in the pedigree with their maximal risk to have an affected child with CMT given that III-8 and IV-1 through IV-9 are under age 25.

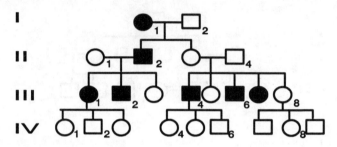

(A) 100 percent
(B) 50 percent
(C) 25 percent
(D) 12.5 percent
(E) Virtually 0

215. Individual II-3

216. Individual III-1

217. Individual III-3

218. Individual III-8

219. Individual IV-8

Questions 220–224

The pedigree in the figure below shows two members of a sibship affected with Charcot-Marie-Tooth disease. Match the following individuals with the probability that their child will be affected. Assume that the incidence of this type of CMT in the general population is 1 in 10,000.

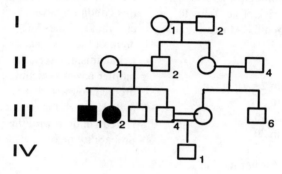

(A) 1 in 24
(B) 1 in 100
(C) 1 in 300
(D) 1 in 400
(E) 1 in 800

220. Individual II-3

221. Individual III-1

222. Individual III-3

223. Individuals III-4 or III-5

224. Individual III-6

Questions 225–229

A woman who has two brothers with hemophilia A, an X-linked recessive disorder, and has had two normal sons is again pregnant. She requests counseling for the risk of her fetus to have hemophilia. Match the following risks.

(A) 1
(B) 1 in 2
(C) 1 in 5
(D) 1 in 10
(E) 1 in 20

225. Risk for her mother to be a carrier

226. Risk for her to be a carrier if she had no children

227. Risk for her to be a carrier given two normal sons

228. Risk for her to have an affected child

229. Risk for her child to have hemophilia given it is a male

Questions 230–233

Waardenburg syndrome is an autosomal dominant condition that accounts for 1.4 percent of congenitally deaf persons. In addition to deafness, patients with this condition have a typical facies that includes lateral displacement of the inner canthi and partial albinism. Given that the mother has Waardenburg syndrome and the father is unaffected, match the number of affected children with the probability below.

(A) 1/8
(B) 1/4
(C) 1/3
(D) 3/8
(E) 1/2

230. None of three children affected

231. One of three children affected

232. Two of three children affected

233. Three of three children affected

Mendelian Inheritance

Answers

138. The answer is C. *(Gelehrter, pp 29–36. Thompson, 5/e. pp 59–66.)* Autosomal dominant conditions tend to have a vertical pattern in the pedigree (see figure in question group 179–183, Pedigree I). Males and females are affected with equal frequency and severity. The frequency of isolated cases (presumably due to new mutation) increases with the severity of the disorder; that is, the lesser the reproductive fitness, the more likely it is to have resulted from a new mutation. Autosomal dominant conditions are generally clinically variable and pleiotropic (multiple phenotypic features that frequently appear to be unrelated).

139. The answer is A. *(Gelehrter, pp 36–39. Thompson, 5/e. pp 66–71.)* Autosomal recessive conditions tend to have a horizontal pattern in the pedigree. Males and females are affected with equal frequency and severity. It is the pattern of inheritance most often seen in cases of deficient enzyme activity (inborn errors of metabolism). Autosomal recessive conditions tend to be more severe than dominant conditions and are less variable than dominant phenotypes. Both alleles are defective but do not necessarily contain exactly the same mutation. All persons carry 6 to 12 mutant recessive alleles. Fortunately, most matings involve persons who carry mutations at different loci. Since related persons are more likely to inherit the same mutant gene, consanguinity increases the possibility of homozygous affected offspring.

140. The answer is D. *(Gelehrter, pp 39–44. Thompson, 5/e. pp 74–82.)* In an X-linked dominant disorder, affected males transmit their single affected X-chromosome to all their daughters. The unaffected chromosome is the Y chromosome, which is transmitted to all sons. In the Turner (XO) phenotype, females have only one X-chromosome and therefore may not have inherited the affected allele. In some cases, favorable lyonization will result in the inactivation of the majority of X chromosomes that bear the affected allele. Back mutation may result in the reappearance of the normal allele. Nonpaternity would also explain the lack of inheritance of the abnormal phenotype.

141–148. The answers are: 141-C, 142-D, 143-E, 144-A, 145-A, 146-C, 147-B, 148-B. *(Gelehrter, pp 22–46. Thompson, 5/e. pp 53–88.)* Males must transmit their Y chromosome to produce sons, which rules out the possibility for male-to-male transmission of X chromosome alleles. Males transmit their X chromosome to their daughters, so that female offspring of males with X-linked disease are obligate carriers. Elevated paternal age (new mutations) and variable expressivity are characteristics of autosomal dominant inheritance, while inborn errors of metabolism and consanguinity are associated with autosomal recessive inheritance. Elevated maternal age is associated with an increased risk for chromosomal nondisjunction. An increasing risk of recurrence according to the number of relatives affected is a characteristic of polygenic inheritance.

149–153. The answers are: 149-A, 150-B, 151-A, 152-C, 153-C. *(Gelehrter, pp 39–44. Thompson, 5/e. pp 74–82.)* In X-linked inheritance, no male-to-male transmission of the phenotype occurs and unaffected males do not transmit the phenotype. When the disorder is genetically lethal in males (i.e., the affected male does not reproduce), approximately one-third of cases arise as the result of a new mutation; the remaining two-thirds have heterozygous carrier mothers who generally are unaffected. This phenomenon is known as the *Haldane hypothesis*. When the disorder is lethal to males in utero, only females will be affected and women will have more daughters than sons. When the disorder is nearly always manifest in females, there will be about twice as many affected females as males (since they have twice the chance of receiving an X chromosome).

154–158. The answers are: 154-A, 155-C, 156-B, 157-E, 158-D. *(Gelehrter, p 27. Thompson, 5/e. pp 53–54.)* A person who is homozygous has the same allele on each chromosome at a given locus. A heterozygote or, in the case of an autosomal recessive disorder, a carrier has one normal allele and one mutant allele at a given locus. A compound heterozygote has two different mutant alleles and a double heterozygote has one mutant allele at each of two different loci. The term *hemizygote* refers to an X-linked gene. Since males have only one X chromosome and, therefore, only one allele at any locus on the X chromosome, they are said to be hemizygous.

159–166. The answers are: 159-A, 160-A, 161-D, 162-B, 163-C, 164-B, 165-B, 166-A. *(Thompson, 5/e. pp 58–59.)* The member of the family who first brings that family to medical attention is called the *proband,* or *propositus*; if that person is affected, he or she is called the *index case.* Other relatives may then be classified according to the number of meioses between the two persons (and therefore the likely number of genes in common). First-degree relatives include parents, siblings, and offspring. Second-degree relatives include grandparents and grandchildren, aunts and uncles (regardless of which side of the family), nephews and nieces, and half-siblings. Third-degree relatives include first cousins. Unless the union is consanguineous, spouses are genetically unrelated.

167–170. The answers are: 167-C, 168-B, 169-B, 170-C. *(Gelehrter, pp 27–47. Thompson, 5/e. pp 53–88.)* Diploid persons have two alleles per autosomal locus with one being transmitted to each gamete (Mendel's law of segregation). The key to blood group problems is to recognize that a blood type is ambiguous regarding possible alleles—type A persons may have AA or AO genotypes. Once the possible genotypes are deduced from the blood types, potential offspring will represent all combinations of parental alleles.

171–173. The answers are: 171-D, 172-B, 173-A. *(Gelehrter, pp 36–39. Thompson, 5/e. pp 66–72.)* If the abnormal allele is represented as p and the normal as P, an infant affected with PKU will have the genotype pp. Parents must be heterozygotes or carriers (Pp) for the child to inherit the p allele from both mother and father (assuming correct paternity and the absence of unusual chromosomal segregation). Subsequent children have a 1 in 2 chance of inheriting allele p from the mother and a 1 in 2 chance of inheriting allele p from the father; the chance that both events will occur to give genotype pp is thus $1/2 \times 1/2 = 1/4$, or 25 percent. A normal sibling may be genotype PP (1 in 4 probability) or Pp (1 in 2 probability since two different combinations of parental alleles give this genotype). The ratio of these probabilities results in a 2 in 3 chance (67 percent) of genotype Pp (note that genotype pp is excluded because a normal sibling was specified).

174–178. The answers are: 174-D, 175-B, 176-B, 177-E, 178-A. *(Gelehrter, pp 27–47. Thompson, 5/e. pp 53–88.)* The figure below shows the correctly drawn pedigree with generations indicated by Roman numerals and individuals by arabic numbers.

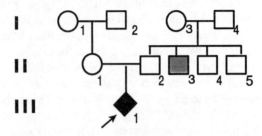

The person who prompted genetic concern is the proband (III-1). Individual II-5, like George, has a brother with cystic fibrosis. Since the parents (I-3, I-4) are carriers, the brother had a 1 in 4 chance of being normal, a 2 in 4 chance of being a carrier, and a 1 in 4 chance of being affected with cystic fibrosis. If the possibility of being affected is eliminated (as given in the pedigree), the odds of being a carrier are 2 in 3.

179–183. The answers are: 179-A, 180-A, 181-B, 182-C, 183-D. *(Gelehrter, pp 27–47. Thompson, 5/e. pp 53–88.)* Pedigree I has the vertical pattern suggestive of autosomal dominant inheritance. Although all affected individuals are male, X-linked inheritance is ruled out by the instance of male-to-male transmission. Affected individuals will have a 50 percent chance of affected offspring, regardless of sex. Pedigree II has the horizontal pattern of autosomal recessive inheritance and this is supported further by the presence of consanguinity. The proband has a 2 in 3 chance of being a carrier, and the consanguinity suggests that his wife is homozygous for the same RP alleles. The risk is thus 2/3 × 1/2 for the proband to contribute the recessive allele, while his wife has a 100 percent chance to do so, which results in a final probability of 1 in 3 that the child will be affected. This decreases to 1 in 9 if the proband marries his wife's unaffected sister, who also has a 2 in 3 chance of being a carrier.

184–187. The answers are: 184-C, 185-E, 186-E, 187-C. *(Gelehrter, pp 27–47. Thompson, 5/e. pp 53–88.)* Pedigree III has the oblique pattern of X-linked recessive inheritance. The proband has a 1 in 2 chance of being a carrier, and thus each of her sons has a 1 in 4 chance of being affected (but overall there is a 1 in 8 chance of an affected child since 1 in 2 of her children will be male). Unless there is nonrandom lyonization of the X chromosome that carries the normal RP allele, female carriers should not be affected. X-linked inheritance implies that males cannot transmit diseases to their sons and that their daughters will be obligate carriers.

188–190. The answers are: 188-B, 189-B, 190-C. *(Gelehrter, pp 36–39. Thompson, 5/e. pp 66–72.)* Representing the abnormal allele as A and the normal as a, achondroplastic parents (Aa) will produce AA, Aa, and aa children with respective probabilities of 1 in 4, 2 in 4, and 1 in 4. Since both AA and Aa children will be affected with achondroplasia, the total risk is 3 in 4 (75 percent) for an achondroplastic child. Chance has no memory, so the same risk applies for each pregnancy regardless of prior outcomes. If an achondroplast (Aa) and normal person (aa) marry, their risk is 50 percent to have a child with achondroplastic dwarfism (Aa). The homozygous individual with achondroplasia (genotype AA) is in fact more severely affected and often dies in the newborn period.

191–195. The answers are: 191-D, 192-C, 193-E, 194-A, 195-E. *(Gelehrter, pp 39–45. Thompson, 5/e. pp 74–83.)* Women have two X chromosomes; men an X chromosome and a Y chromosome. X-linked inheritance implies that the abnormal allele resides on the X chromosome. Males have one allele that is either normal or abnormal (hemizygosity), while women will usually be homozygous normal or heterozygous (carriers). Rarely, women may be homozygous abnormal. Female offspring of affected males must receive their father's X to be female; they are thus obligate carriers of the abnormal allele. Carrier females have a 25 percent chance to have affected children, or a 50 percent chance given that the offspring is male. Affected males cannot transmit the disease to sons because, by definition, they have transmitted their Y rather than their X chromosome.

196–199. The answers are: 196-D, 197-E, 198-A, 199-B. *(Gelehrter, pp 27–45. Thompson, 5/e. pp 53–95.)* The Lyon hypothesis postulates early and irreversible inactivation of X chromosomes in female embryos and states that the inactivation process is random with regard to which X is chosen in each primordial tissue. In rare female carriers with nonrandom X-inactivation, the X with an abnormal allele remains active and produces symptoms of the X-linked disease. Autosomal dominant diseases often vary in severity within families (variable expressivity), but occasionally are clinically silent in a person known to carry the abnormal allele (incomplete penetrance). Normal parents can have one child with an autosomal dominant condition via mutation, but two affected children must mean that one parent has a germ cell lineage with the mutation. Albinism is one of many genetic diseases that exhibit genetic heterogeneity, which means that mutations at several loci can produce identical phenotypes.

200–204. The answers are: 200-A, 201-B, 202-D, 203-C, 204-C.
(Gelehrter, pp 36–39. Thompson, 5/e. pp 66–72.) Parents of children with autosomal recessive disorders are obligate carriers if nonpaternity and rare examples of uniparental disomy (inheritance of chromosomal homologues from the same parent) are excluded. Normal siblings have a 2 in 3 chance to be carriers since they cannot be homozygous for the abnormal allele. Grandparents have a 1 in 2 chance to be carriers because one or the other of them must have transmitted the abnormal allele to the obligate carrier parent. First cousins share a set of grandparents of whom one must be a carrier. There is a 1 in 2 chance for the aunt or uncle to be a carrier and a 1 in 4 chance for the first cousin. Half-siblings share an obligate carrier parent and have a 1 in 2 chance to be carriers. These calculations assume the lack of mutations (Tay-Sachs is rare) and the lack of coincidental alleles (no consanguinity).

205–209. The answers are: 205-C, 206-D, 207-E, 208-B, 209-E.
(Gelehrter, pp 27–45. Thompson, 5/e. pp 53–95.) Parents of a child affected with an autosomal recessive disease will usually be carriers with a 1 in 4 risk to have a second affected child. If a son is affected with an X-linked disease, the mother may be a carrier (1 in 4 risk for subsequent children) or not (virtually 0 risk). Children affected with autosomal dominant diseases either represent a new mutation (neither parent affected with virtually 0 recurrence risk) or inherit their abnormal allele from a parent (50 percent recurrence risk). If the parents are phenotypically normal, they will have virtually 0 recurrence risk unless one has the abnormal allele (incomplete penetrance). In the latter case, their recurrence risk is 50 percent. For X-linked dominant inheritance, the affected male or female may represent a new mutation (virtually 0 recurrence risk) or transmission from an affected mother (50 percent recurrence risk).

210–214. The answers are: 210-E, 211-B, 212-C, 213-C, 214-D.
(Gelehrter, pp 39–44. Thompson, 5/e. pp 72–82.) The predominance of affected males with transmission through females makes this pedigree diagnostic of X-linked recessive inheritance. Individual I-1 is an obligate carrier as demonstrated by her affected son and grandson. Individual II-2 cannot transmit an X-linked disorder, although his daughters are obligate carriers. Individual II-3 must be a carrier because of her affected son, which results in a 1 in 4 probability for recurrence of CMT

in her offspring. Individual II-5 has a 1 in 2 probability of being a carrier with a 1 in 8 probability for affected offspring. Individual III-4 also has a 1 in 2 probability of being a carrier; her risk for affected offspring is also 1 in 8 despite the consanguineous marriage. Individual III-8 has a 1 in 4 chance of being a carrier and a 1 in 16 chance of having affected offspring.

215–219. The answers are: 215-B, 216-B, 217-E, 218-C, 219-D. *(Gelehrter, pp 29–36. Thompson, 5/e. pp 59–66.)* The vertical inheritance pattern (multiple affected generations) and equal sex ratio is suggestive of autosomal dominant inheritance for this form of CMT disease. Individual III-1 is affected and has a 50 percent recurrence risk in each offspring. Individual II-3 is most likely an example of incomplete penetrance since she has three affected children. Individual III-3 is old enough to show the disease but is not affected and thus has a virtually 0 recurrence risk (excluding the possibility of incomplete penetrance). Individual III-8 is not old enough to manifest CMT and therefore has a maximal 50 percent chance to have the abnormal allele with a maximal 25 percent chance for affected children. Her daughter (individual IV-8) thus has a maximal 12.5 percent chance that her child will be affected.

220–224. The answers are: 220-D, 221-B, 222-C, 223-A, 224-E. *(Gelehrter, pp 36–39. Thompson, 5/e. pp 66–72.)* Affected siblings of different sexes suggest autosomal recessive inheritance. Individual III-1 will thus be homozygous for the abnormal CMT allele, while the probability that his future wife will be a carrier is twice the square root of the population incidence (2 pq from the Hardy-Weinberg law = 2 × 1 × 100). The probability that his child will be affected is thus 1/50 × 1/2 = 1/100. Similarly, individual III-3 has a 2 in 3 chance of being a carrier and a 1 in 300 chance that his child will be affected. Individual III-4 has a 2 in 3 chance of being a carrier, and III-5 a 1 in 4 chance based on their common grandparents. The chance their second child will be affected is 2/3 × 1/4 × 1/4 = 1/24. Individual III-6 also has a 1 in 4 chance to be a carrier and the chance that his child will be affected is 1/4 × 1/50 × 1/4 = 1/800. For II-3, her chance of being a carrier is 1 in 2, the chance for an unrelated husband to be a carrier 1 in 50, and the final probability for an affected child is 1/2 × 1/50 × 1/4 = 1/400.

225–229. The answers are: 225-A, 226-B, 227-C, 228-E, 229-D. *(Gelehrter, pp 264–267. Thompson, 5/e. pp 400–409.)* Since the consultand's mother has two affected sons with hemophilia, she is an obligate carrier (excluding rare instances of germinal mosaicism). Her daughter has a 1 in 2 chance to receive the X that carries the abnormal gene.

However; the fact that the daughter has two normal boys can be taken into account by using Bayes theorem. The easiest way to do this is to set up a table as shown below:

	Woman Is Carrier	Not Carrier
Prior probability	1/2	1/2
Conditional (given 2 normal boys)	1/4	1
Joint (prior × conditional)	1/8	4/8
Adjusted probability	$\dfrac{1/8}{1/8 + 4/8} = 1/5$	$\dfrac{4/8}{1/8 + 4/8} = 4/5$

Given a 1 in 5 chance to be a carrier, the woman's risk for an affected child is $1/5 \times 1/4 = 1/20$ (1/10 for a male to be affected).

230–233. The answers are: 230-A, 231-D, 232-D, 233-A. *(Gelehrter, pp 30–31. Thompson, 5/e. pp 146–147.)* For each pregnancy, the probability that the child will be affected is 1 in 2. Therefore the probability that all three children will be affected is the product of the three independent events, that is, $1/2 \times 1/2 \times 1/2 = 1/8$. The probability that all three children will be unaffected is the same. When evaluating the probability that one of the three children will be affected, it must be noted that there are three of eight possible birth orders that have one affected child (Www, wWw, wwW). The probability of two affected children is also 3 in 8. In general, one can determine the probability of any given number of affected children by using the binomial expansion $(p + q)^n$, where p and q are the probabilities of two alternative events $(1 - p = q)$ and n is the number of events. In this case, $p = 1/2$, $q = 1/2$, and $(p + q)^3 = p^3 + 3p^2q + 3pq^2 + q^3$.

$$p^3 = 1/8 = \text{probability of 3 unaffected}$$
$$3p^2q = 3/8 = \text{probability of 2 unaffected, 1 affected}$$
$$3pq^2 = 3/8 = \text{probability of 1 unaffected, 2 affected}$$
$$q^3 = 1/8 = \text{probability of 3 affected}$$

Population Genetics and Polygenic Inheritance

DIRECTIONS: Each question below contains five suggested responses. Select the **one best** response to each question.

234. In a population in which random mating occurs, 50 percent of the people in one generation have the genotype Aa. In the following generation, the percentage of people with genotype Aa will be

(A) 100 percent
(B) 75 percent
(C) 50 percent
(D) 25 percent
(E) 0 percent

235. A GT (guanidine thymidine) polymorphism has five different possible alleles, each with a frequency of 0.2. The percentage of people who are heterozygous is

(A) 20 percent
(B) 40 percent
(C) 50 percent
(D) 60 percent
(E) 80 percent

236. All the following assumptions must be valid for the Hardy-Weinberg equilibrium to apply to allele frequencies in a population EXCEPT

(A) no selection
(B) random mating
(C) no mutation
(D) no expansion
(E) no migration

237. Galactosemia is an inborn error of metabolism in which infants present with failure to thrive, vomiting, jaundice, hepatomegaly, and cataracts. The frequency of this disorder is approximately 1/40,000 live births (0.000025). The frequency of the carrier state is approximately

(A) .02
(B) .01
(C) .005
(D) .002
(E) .001

238. For recessive diseases, the Hardy-Weinberg term $p^2 + 2pq + q^2$ provides the explanation for three important phenomena. First, elimination of homozygous abnormal individuals will not significantly decrease the incidence of the disease. Second, an advantage for heterozygous carriers may have a dramatic effect in increasing the incidence of a recessive disease. Third, parental consanguinity should alert the physician to the possibility of recessive disease in offspring. Recalling that q is usually employed for the frequency of abnormal alleles in recessive disorders, select the explanation of these phenomena.

(A) $q^2 > p^2$
(B) $2pq > p^2$
(C) $q^2 > 2pq$ and $2pq > p^2$
(D) $p^2 > 2pq > q^2$
(E) $p^2 > q^2$

239. Achondroplasia is an autosomal dominant form of skeletal dysplasia that produces dwarfism. Rarely, two affected individuals (heterozygotes) will mate and produce a severely affected homozygote. Representing the abnormal allele frequency by p and the normal allele frequency by q, why does the Hardy-Weinberg law predict that homozygotes will be rare in dominant diseases?

(A) Affected individuals, represented by the genotype frequency q^2, will be very rare
(B) Affected individuals, represented by the genotype frequency p^2, will be very rare
(C) The genotype frequency $2pq$, which represents affected heterozygotes, will be much larger than p^2, which represents affected homozygotes
(D) Assortative mating is rare because achondroplasts do not meet
(E) Achondroplasia in homozygotes is prenatally lethal

240. Achondroplasts have about 80 percent less viable offspring than do normal persons. This has certain implications for the mutation rate since the incidence of achondroplasia is thought to have remained constant for some time. Representing fitness by f, the coefficient of selection by s, and the mutation rate by μ, a true statement is

(A) s is 0.8, f is 0.2, μ is relatively high
(B) s is 0.4, f is 0.2, μ is relatively low
(C) s is 0.2, f is 0.8, μ is relatively high
(D) s is 0.2, f is 0.2, μ is relatively high
(E) s is 0.2, f is 0.8, μ is relatively low

241. Isolated cleft lip and palate is a multifactorial trait. The recurrence risk of isolated cleft lip and palate is

(A) the same in all families
(B) not dependent upon the number of affected family members
(C) the same in all ethnic groups
(D) the same in males and females
(E) affected by the severity of the cleft

Questions 242–244

A newborn boy dies with severe hyperammonemia. Prior to his death, a diagnosis is made of ornithine transcarbamylase (OTC) deficiency, an X-linked disorder of urea cycle metabolism, which in its classic form is lethal in males. Answer the following questions with regard to this family.

242. The likelihood that the proband's disease is the result of a new mutation is

(A) 100 percent
(B) 67 percent
(C) 50 percent
(D) 33 percent
(E) 25 percent

243. Subsequent testing reveals that the proband's mother is a carrier of the disease. The chance that her next child will be affected is

(A) 67 percent
(B) 50 percent
(C) 33 percent
(D) 25 percent
(E) 0 percent

244. The proband's mother remarries. The chance that her next child will be affected is

(A) 67 percent
(B) 50 percent
(C) 33 percent
(D) 25 percent
(E) 0 percent

245. All the following statements regarding twins are true EXCEPT

(A) the frequency of monozygotic twins is greater than the frequency of dizygotic twins
(B) the frequency of twin births is different in different ethnic groups
(C) among North American whites, the incidence of twinning is approximately 1/87 live births
(D) monozygotic twins may be mono- or dichorionic
(E) 50 percent of dizygotic twin pairs are same sex

246. Many disorders that present in adult life, such as coronary artery disease and hypertension, are multifactorial traits. A multifactorial trait results from

(A) the interaction between the environment and a single gene
(B) the interaction between the environment and multiple genes
(C) multiple postnatal environmental factors
(D) multiple pre- and postnatal environmental factors
(E) multiple genes independent of environmental factors

DIRECTIONS: Each group of questions below consists of lettered headings followed by a set of numbered items. For each numbered item select the **one** lettered heading with which it is **most** closely associated. Each lettered heading may be used **once, more than once, or not at all.**

Questions 247–251

Match the following.

(A) Quantitative trait
(B) Random fluctuation
(C) Reproductive success
(D) Increased incidence with polymorphism
(E) Ancestral mutation

247. Founder effect

248. Fitness

249. Genetic drift

250. Polygenic inheritance

251. Heterozygosity

Questions 252–253

In a population in Hardy-Weinberg equilibrium in which 16 percent of the people have genotype AA, match the genotypes below with their percentage of the population (assume a and A are the only alleles possible).

(A) 16 percent
(B) 24 percent
(C) 36 percent
(D) 48 percent
(E) 84 percent

252. Genotype aa

253. Genotype Aa

Questions 254–258

A survey is performed by the blood bank to determine donor availability. Match the blood type data on surveyed individuals with the allele frequencies they would imply for the populace. Assume that type A or B individuals are not homozygous.

(A) Allele O = 0.88, allele A = 0.10, allele B = 0.02
(B) Allele O = 0.75, allele A = 0.17, allele B = 0.08
(C) Allele O = 0.78, allele A = 0.15, allele B = 0.07
(D) Allele M = 0.50, allele N = 0.50
(E) Allele M = 0.60, allele N = 0.40

254. 30 type O individuals, 13 type A, 5 type B, 2 type AB

255. 60 type O individuals, 26 type A, 10 type B, 4 type AB

256. 19 type O individuals, 5 type A, 1 type B

257. 15 type M individuals, 30 type MN, 5 type N

258. 10 type M individuals, 20 type MN, 10 type N

Questions 259–260

DNA analysis of a two-allele polymorphism (A,a) is performed on a large population of unrelated individuals. Twenty-five percent of people are found to be homo-zygous (AA). Match the alleles below with their frequencies.

    (A) .125
    (B) .25
    (C) .5
    (D) .625
    (E) .75

259. A

260. a

Questions 261–263

Assume that frequencies for the different blood group alleles are as follows: A = .3; B = .1; and O = .6. Match the following blood types with the percentage of people expected to have those blood types.

    (A) 7 percent
    (B) 13 percent
    (C) 27 percent
    (D) 36 percent
    (E) 45 percent

261. Blood type A

262. Blood type B

263. Blood type O

Questions 264–268

Based on frequencies of 0.7 for allele O, 0.2 for allele A, 0.1 for allele B, 0.6 for allele M, and 0.4 for allele N in a population sample, match the indicated blood types with their expected number per 1000 people. The ABO and MN blood group loci are not linked.

    (A) 490
    (B) 154
    (C) 150
    (D) 40
    (E) 24

264. Type AB

265. Type O

266. Type B

267. Type A, MN

268. Type B, N

Questions 269–275

Screening of a West African population for sickle cell disease reveals 16 individuals with homozygous sickle cell anemia (genotype SS), 1600 with sickle trait (genotype AS), and 4784 normal individuals (genotype AA). Match the allele and genotype figures below.

    (A) 0.125
    (B) 100
    (C) 1.1
    (D) 0.16
    (E) None of above

269. Frequency of allele S

270. Frequency of allele A

271. Expected number of SS individuals if Hardy-Weinberg equilibrium is obeyed

272. Expected number of AS individuals if Hardy-Weinberg equilibrium is obeyed

273. Estimated fitness of SS genotype

274. Estimated fitness of AS genotype

275. Estimated fitness of AA genotype

Questions 276–278

Tay-Sachs disease is a recessive degenerative neurologic disorder. The frequency of Tay-Sachs carriers among Ashkenazi Jews is 1/30. The frequency of Tay-Sachs carriers among whites of Western European descent is approximately 1/300. Using this information, choose the odds that a child of the following unions will have Tay-Sachs disease.

    (A) 1/120
    (B) 1/240
    (C) 1/3600
    (D) 1/9000
    (E) 1/36,000

276. Both parents are Ashkenazi Jews

277. The mother has an affected child by a previous union; the father is an Ashkenazi Jew

278. The mother is an Ashkenazi Jew; the father is a white Western European

Questions 279–283

Many common birth defects such as cleft palate or myelomeningocele follow the polygenic, or multifactorial, inheritance model with threshold. This model attempts to explain the following empirical risks that a relative faces if there is an affected person in the family: identical twin 20 to 40 percent, first-degree relative 2 to 3 percent, two first-degree relatives 4 to 6 percent, three first-degree relatives 6 to 9 percent, second-degree relative 0.5 percent, and so on. Based on these figures, match the individuals below with their risk.

   (A) 20 to 40 percent
   (B) 6 to 9 percent
   (C) 4 to 6 percent
   (D) 2 to 3 percent
   (E) 0.5 percent

279. Child of a parent with spina bifida

280. Grandchild of a parent with cleft palate

281. Twin brother of a girl with congenital dislocated hip

282. Sibling of two children with cleft palate and normal parents

283. Sibling of two children with a congenital heart defect and an affected parent

Population Genetics and Polygenic Inheritance

Answers

234. The answer is C. *(Gelehrter, pp 49–52. Thompson, 5/e. pp 143–152.)* When random mating occurs, the frequency of genotypes remains constant over time. Therefore, if the frequency of the genotype Aa is 50 percent in one generation, it will remain 50 percent in subsequent generations. This situation is known as *Hardy-Weinberg equilibrium*.

235. The answer is E. *(Gelehrter, pp 146–147. Thompson, 5/e. pp 143–152.)* For each allele, the frequency of homozygotes is $(0.2)^2$. Therefore, the total number of homozygotes in this five-allele system is $5(0.2)^2$, or 0.2. The frequency of heterozygotes is $1 -$ homozygotes, or 0.8. The answer can also be calculated using the binomial expansion $(p + q)^5$.

236. The answer is D. *(Gelehrter, pp 49–65. Thompson, 5/e. pp 143–165.)* For a single locus with alleles M and N, the frequency of allele M can be represented by p and that of allele N by q. Obviously, $p + q = 1$. Under certain conditions, as pointed out independently by Hardy and Weinberg, the proportions of individuals with genotypes MM, MN, and NN will derive from the expansion $(p + q)(p + q) = p^2 + 2pq + q^2$. If the frequency of M is 0.9 and that of N is 0.1, then the proportions of MM (0.81), MN (.18), and NN (0.01) individuals would be predicted. Note that these proportions ($p^2 + 2pq + q^2$) must also add up to 1. If, however, the measured proportions were 0.9 for MM, 0.1 for MN, and virtually 0 for NN, then there would be deviation from the Hardy-Weinberg law. Such discrepancies occur when there is selection (e.g., lethality of genotype NN), migration, or nonrandom mating (e.g., mutual aversion of MN heterozygotes). Population expansion could occur without altering the Hardy-Weinberg equilibrium.

237. The answer is B. *(Gelehrter, pp 49–52. Thompson, 5/e. pp 145–153.)* The Hardy-Weinberg expansion, $p^2 + 2pq + q^2$, describes the frequency of genotypes for allele frequencies p and q. In the case of rare disorders ($q^2 < 0.0001$), p approaches 1. The heterozygote frequency $2pq$ is thus approximately $2q$. In this case, $q^2 = 0.000025$, $q = 0.005$, and $2q = .01$.

238. The answer is D. *(Gelehrter, pp 49–65. Thompson, 5/e. pp 143–165.)* In a recessive disease, the majority of abnormal alleles will reside in carriers since $p > q$ and $p^2 > 2pq > q^2$. For example, a disorder of incidence 1/10,000 (quite common for a recessive disease) will have $q = 1/100$, $p = 99/100$, and $2pq = 1/50$. In a population of 10 million, there will be 9.8 million normal individuals, 200,000 carriers, and 1000 affected individuals for this disease locus (assuming minimal selection). Eliminating the 1000 affected individuals will thus do little to impact the 200,000 abnormal alleles in carriers, but biologic or political factors that encourage carrier reproduction will have a 200-fold greater impact on abnormal allele frequency than measures that enhance homozygote survival. Since carriers are still quite rare compared with normal individuals, matching of rare recessive alleles is greatly enhanced when there is common descent through consanguinity.

239. The answer is C. *(Gelehrter, pp 49–65. Thompson, 5/e. pp 143–165.)* The Hardy-Weinberg term $p^2 + 2pq + q^2$ is useful for considering relative frequencies of genotypes even though these will be modified slightly by selection, migration, or inbreeding in actual populations. Since the incidence of achondroplasia is less than 1 in 10,000 births, the frequency of the abnormal allele (usually p is taken as the abnormal allele frequency in dominant diseases) is quite small (q is approximately equal to 1). For this reason, $p^2 < 2pq < q^2$ and homozygous abnormal patients will be extremely rare for most dominant diseases. Assortative mating (preferential mating among certain genotypes) is not uncommon in achondroplasia because of activities sponsored by organizations such as Little People of America. Such matings are the main source of homozygotes.

240. The answer is A. *(Gelehrter, pp 49–65. Thompson, 5/e. pp 143–165.)* If an abnormal allele is as likely to be transmitted to the next generation as its corresponding normal allele, it is said to have a fitness (f) of 1. Loss of fitness (decrease in allele frequency after one generation) is measured by the coefficient of selection (s) where $f = 1 - s$. If achondroplasts have only 1/5 the children of parents of normal stature in the population, then their fitness is 0.2 and the coefficient of selection is 0.8. The achondroplast alleles eliminated by selection must be replaced by mutation if the disorder has not disappeared or declined in incidence. Thus the mutation rate (μ) would be expected to be high relative to more benign dominant diseases.

241. The answer is E. *(Gelehrter, pp 57–65. Thompson, 5/e. pp 349–363.)* Cleft lip with or without cleft palate—CL(P)—is one of the most common congenital malformations. Because of the genetic component of this trait, it tends to be more common in certain families. The more family members affected and the more severe the cleft, the higher the recurrence risk. In addition, CL(P) is more common in males and in certain ethnic groups (Asians > whites > American blacks).

242–244. The answers are: 242-D, 243-D, 244-D. *(Gelehrter, pp 39–44. Thompson, 5/e. pp 72–82, 163.)* For X-linked lethal disorders, two-thirds of the cases result from an abnormal gene passed on by a carrier mother and one-third of cases result from a new mutation. This rule is known as the Haldane hypothesis and is named for J.B.S. Haldane who first described it in 1935. A woman who is a carrier of an X-linked lethal disorder will pass that gene on to half her children, with the following results: 1/4 carrier daughters, 1/4 affected sons, 1/2 unaffected children. Since this is an X-linked disorder, these proportions are not dependent on the father in any way and will not change if the mother remarries.

245. The answer is A. *(Thompson, 5/e. pp 389–391.)* Twins may be either monozygotic (identical) or dizygotic (fraternal). Monozygotic twins arise from a single zygote, which divides during early embryonic development. Depending on the timing of that division, monozygotic twins may be mono- or diamniotic and mono- or dichorionic. Dizygotic twins arise from two different zygotes and are diamniotic and dichorionic. The frequency of monozygotic twins is the same in all populations (approximately 1 in 260 live births), whereas the frequency of dizygotic twins is different in different populations. The frequency of any type of twinning among North American whites is about 1/87 live births. All monozygotic twins and 50 percent of dizygotic twins are same sex.

246. The answer is B. *(Gelehrter, pp 57–65. Thompson, 5/e. pp 349–363.)* Many common disorders tend to run in families but are not single-gene or chromosomal disorders. These disorders are multifactorial traits caused by multiple genetic and environmental factors. The term *polygenic* may be used to refer to this genetic component.

247–251. The answers are: 247-E, 248-C, 249-B, 250-A, 251-D. *(Thompson, 5/e. pp 143–165.)* In a small population, whether because of geographic isolation or limited fecundity, the frequency of a mutant allele may vary widely because of unequal sampling of mutant or non-mutant gametes. These variations in allele frequency are not due to reproductive capability (i.e., number of viable descendants equals fitness) but to sampling effects (genetic drift). If the founder of a small population has a certain mutant allele, its frequency may be maintained at unusually high levels (founder effect). Founder effects represent a special type of genetic drift. The occurrence of different alleles at a locus is termed *heterozygosity,* and this will increase with the degree of polymorphism. These multiple alleles may have incremental effects on phenotype and produce continuous variation typical of a quantitative trait. Dependence of a trait on multiple alleles or loci is termed *polygenic inheritance* and is exemplified by blood pressure and height.

252–253. The answers are: 252-C, 253-D. *(Gelehrter, pp 49–52. Thompson, 5/e. pp 143–152.)* The Hardy-Weinberg equation states that the proportion of individuals with a given genotype in a two-allele system can be determined from the binomial expansion $(p + q)^2$ where p and q are the frequencies of the two alleles (and $p + q = 1$). The frequencies of the genotypes are p^2, $2pq$, and q^2. In this case, AA or $p^2 = 0.16$ and A $= 0.4$. Since A $+$ a $= 1$, a $= 0.6$ and aa or $q^2 = 0.36$. The frequency of Aa is $2pq$ or $(2)(0.4)(0.6) = 0.48$. (Note that $0.16 + 0.48 + 0.36 = 1$.)

254–258. The answers are: 254-C, 255-C, 256-A, 257-E, 258-D. *(Gelehrter, pp 49–65.)* Allele frequency is calculated by dividing the number of alleles of a specified type (e.g., allele O) by the total number of alleles (two alleles per individual for a single locus in a diploid population). Type O individuals have OO genotypes and thus will contribute two O alleles per individual, while heterozygous type A or B individuals will contribute one O and one A or B allele per individual. Total ABO alleles in the population sample will equal twice the number of individuals. The accuracy of these allele frequencies will depend on sample size and population homogeneity.

259–260. The answers are: 259-C, 260-C. *(Gelehrter, pp 49–52. Thompson, 5/e. pp 143–152.)* Using the Hardy-Weinberg expansion ($p^2 + 2pq$

+ q^2), p^2 is 0.25 and p, or the frequency of the A allele, is 0.5. Since $1 - p = q$, q is also 0.5.

261–263. The answers are: 261-E, 262-B, 263-D. *(Gelehrter, pp 49–52. Thompson, 5/e. pp 121–123, 145–149.)* It is important to remember that individuals with blood type A can have genotype AA or AO, and individuals with blood type B can have genotype BB or BO. Therefore, the frequency of blood type A is the frequency of homozygotes, i.e., (0.3)(0.3), plus the frequency of heterozygotes, i.e., 2(0.3)(0.6), for a total of 0.45. The frequency of blood type B is (0.1)(0.1) + 2(0.1)(0.6) for a total of 0.13. The frequency of individuals with blood type O is simply the frequency of homozygotes (0.6)(0.6) = 0.36.

264–268. The answers are: 264-D, 265-A, 266-C, 267-B, 268-E. *(Gelehrter, pp 49–65. Thompson, 5/e. pp 143–165.)* Allele frequencies indicate the likelihood for that allele to occur at a locus. The proportion of individuals with a given blood type will be represented by the number of genotypes that yield the blood type multiplied by their likelihood to occur. For example, type B results from genotypes BO, OB, and BB; the probability that alleles B and O will occur together is (0.1)(0.7) = 0.07 and that homozygous B will occur is (0.1)(0.1) = 0.01. Type B individuals will constitute 70 + 70 + 10 = 150 individuals per 1000. Since the ABO and MN blood group loci are not linked, their allele frequencies will be independent of one another. The proportion of type B, N individuals will be the joint probability that type B individuals (0.15) will be homozygous for allele N (0.16) = (0.15)(0.16) = 0.240 = 24 per 1000.

269–275. The answers are: 269-A, 270-E, 271-B, 272-E, 273-D, 274-C, 275-E. *(Thompson, 5/e. pp 157–158.)* Of the 12,800 alleles, 1632/6400 will be S alleles, which represents an allele frequency *(s)* of about 1/8, or 0.125. The normal allele frequency (A) is thus 7/8. The terms $a^2 + 2as + s^2$ give the proportion of individuals expected to have SS genotypes [(1/8)(1/8)(6400) = 100], AS genotypes [(2)(1/8)(7/8)(6400) = 1400], and AA genotypes (4900) if Hardy-Weinberg equilibrium is assumed. Comparison of these expected numbers to those observed yields an approximate fitness of 16/100 = 0.16 for the SS genotype, 1600/1400 = 1.1 for the AS genotype, and 4784/4900 = 0.95 for the AA genotype. These figures support a heterozygote advantage for the AS genotype, which is thought to protect individuals from malarial infection.

276–278. The answers are: 276-C, 277-A, 278-E. *(Gelehrter, pp 30–31, 49–52. Thompson, 5/e. pp 143–152.)* In order to determine the joint probability of two or more independent events, one must determine the product of their separate probabilities. In question 276, both parents have a 1/30 chance of carrying an abnormal gene for Tay-Sachs disease and, for each pregnancy, a 1/2 chance of passing that gene along should they carry it. The probability that all of these four independent events will occur is $(1/30)(1/2)(1/30)(1/2) = 1/3600$. If the mother has had a previously affected child, her risk of being a carrier is now 1. The probability in question 277 is therefore $(1)(1/2)(1/30)(1/2) = 1/120$. In question 278, the joint probability is $(1/30)(1/2)(1/300)(1/2) = 1/36,000$.

279–283. The answers are: 279-D, 280-E, 281-D, 282-C, 283-B. *(Gelehrter, pp 49–65. Thompson, 5/e. pp 143–165.)* For a given individual, parents, siblings, and children represent first-degree relatives; grandparents, aunts, uncles, nephews, nieces, and half-siblings represent second-degree relatives, and first cousins represent third-degree relatives. The terms *first degree, second degree,* and *third degree* also indicate the proportion of genes in common (1/2, 1/4, and 1/8, respectively) and the coefficient of inbreeding (1/4, 1/8, and 1/16, respectively). Similar degrees of relationship also indicate similar risks for common birth defects that follow polygenic inheritance as indicated in the question. In actuality, the risks are somewhat different for various types of common birth defects and may increase or decrease greatly according to the sex of the individual at risk.

Clinical Genetics

DIRECTIONS: Each question below contains five suggested responses. Select the **one best** response to each question.

284. Newborn screening is available for several metabolic disorders including phenylketonuria (PKU), maple syrup urine disease (MSUD), and biotinidase deficiency. Important factors to consider before initiating a newborn screening program include all the following EXCEPT

(A) severity of the disorder
(B) community acceptance
(C) cost
(D) benefit of early detection
(E) availability of prenatal diagnosis for the disorder

285. Factors known to affect teratogenicity include all the following EXCEPT

(A) maternal genotype
(B) paternal exposure
(C) fetal genotype
(D) dosage of agent
(E) timing of exposure to agent

286. A couple brings their two children, a boy and a girl, for evaluation because of excessive height, thin body build, loose joints, arachnodactyly (spider fingers) that enables the maneuver shown in the top figure, and bilateral lens dislocation as shown in the bottom figure. The family history indicates the parents are medically normal, but that the paternal grandfather and greatgrandfather died in their forties with lens dislocations and dissecting aortic aneurysms. Your initial conclusions could reasonably include

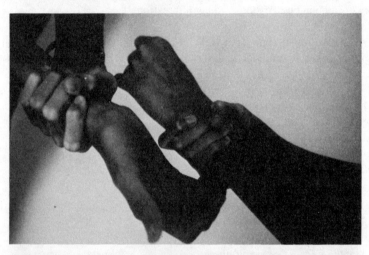

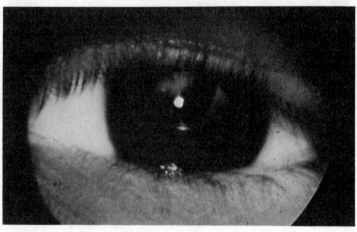

(A) X-linked recessive inheritance
(B) autosomal recessive inheritance
(C) autosomal dominant inheritance with incomplete penetrance
(D) autosomal dominant inheritance with variable expressivity
(E) mitochondrial inheritance

DIRECTIONS: Each group of questions below consists of lettered headings followed by a set of numbered items. For each numbered item select the **one** lettered heading with which it is **most** closely associated. Each lettered heading may be used **once, more than once, or not at all.**

Questions 287–291

Match the following descriptions with the correct term.

(A) Syndrome
(B) Sequence
(C) Disruption
(D) Deformation
(E) Malformation

287. One lesion, serial consequences for related embryonic structures

288. Decreased perfusion because of vascular blow-out (abnormality extrinsic to structure)

289. Crowding because of uterine fibroid tumor or oligohydramnios (abnormality extrinsic to structure)

290. Horseshoe kidney (abnormality intrinsic to structure)

291. One lesion, multiple anomalies affecting unrelated embryonic structures

Questions 292–296

Match the statements below with the type of counseling they exemplify.

(A) Nondirective genetic counseling
(B) Preconceptional counseling
(C) Prenatal counseling
(D) Informative counseling
(E) Supportive counseling

292. We strongly recommend termination of pregnancy because your amniocentesis shows trisomy 21

293. We will tell you your risks; you must make the decision according to your circumstances and family values

294. Your baby has multiple congenital anomalies that will require extensive evaluation to establish a diagnosis

295. Your older son has cystic fibrosis and that implies a 2/3 chance that his sister is a carrier

296. The higher risk for birth defects in pregnancies with uncontrolled diabetes mellitus means that you must establish good management when you decide to attempt pregnancy

Questions 297–301

Cleft lip and palate may be seen as an isolated defect or in association with other findings. Match the following situations below with the correct term.

(A) Sequence
(B) Syndrome
(C) Disruption
(D) Deformation
(E) Malformation

297. Mandibular hypoplasia in utero leads to tongue posteriorly located, which leads to restriction of closure of palatal shelves, which in turn leads to cleft palate

298. Isolated cleft lip and palate in an otherwise normal, healthy infant due to failure of lip fusion and closure of palatal shelves, which results from a multifactorial mode of inheritance

299. Infant with trisomy 13 who has holoprosencephaly, defects of the eye, nose, cleft lip and palate, polydactyly, and scalp defects

300. Mother and child with cleft palate, unusual facies including a flat face and depressed nasal bridge, and myopia

301. Cleft lip and limb reduction following early amniotic rupture and compression of the fetus

Questions 302–306

A man with sickle cell trait (AS genotype) marries a woman with sickle cell trait. Match each potential child with the correct probability.

(A) 1/2
(B) 1/4
(C) 1/8
(D) 1/20
(E) 1/40

302. Normal AA child

303. Child with sickle cell trait (AS)

304. Child with sickle cell anemia (SS)

305. Grandchild with sickle cell anemia (SS), if their AS child marries an unrelated person and the S allele frequency in their population is 0.10

306. Grandchild with sickle cell anemia, if they have two AS children who mate

Questions 307–311

A 2-year-old child presented with the complaint that her urine turned black on the bedsheets after she had enuresis (bedwetting). The parents were aware of no dietary changes and had a negative family history except for the fact that they were first cousins. You collect a urine sample from the child and from a parent; the child's turns relatively dark (left in figure below) after incubation at room temperature. Given the assumption that this is a genetic disease with an abnormal allele (a) frequency of 1/100, match the following risks.

   (A) 1/4
   (B) 1/8
   (C) 1/12
   (D) 1/100
   (E) 1/300

307. Parents to have another affected child

308. Proband to transmit disease with unrelated spouse

309. Proband to transmit disease if she marries her uncle

310. Sibling of proband to transmit disease with an unrelated spouse

311. Sibling of proband to transmit disease if she marries her uncle

Questions 312–316

The metabolite that accumulates in alkaptonuria and is responsible for dark urine is homogentisic acid (HA). The pathway for HA synthesis begins with dietary phenylalanine (PA) that is converted successively by enzymes E1 through E4 to tyrosine (Y), p-hydroxyphenylpyruvate (PHP), HA, and maleylacetoacetic acid (MA).

$$\text{diet} \rightarrow \underset{\text{intestine}}{\text{PA}} \underset{\text{E1}}{\rightarrow} \underset{\text{E2}}{\text{Y}} \rightarrow \underset{\text{E3}}{\text{PHP}} \rightarrow \underset{\text{E4}}{\text{HA}} \rightarrow \text{MA}$$

Match the statements below about this pathway with their likely consequences.

(A) No effect because of enzyme reserve
(B) Accumulation of all intermediates depending on regulation
(C) Deficiency of all intermediates depending on regulation
(D) Accumulation of HA, deficiency of MA
(E) Accumulation of PA, deficiency of Y

312. Deficiency of enzyme E4 as in alkaptonuria

313. Deficiency of enzyme E1

314. Fifty percent activity of enzyme E4 in most tissues

315. Dietary restriction of PA

316. Dietary excess of PA

Questions 317–321

The proband from the pedigree shown in the figure below consults you about her family history of a bleeding disorder. She has medical records that document a diagnosis of von Willebrand's disease (vWD) in her affected father and uncle. The most common mode of inheritance of vWD is autosomal dominant. Based on your own conclusions about the mode of inheritance, match the risks below.

    (A) 100 percent
    (B) 75 percent
    (C) 50 percent
    (D) 25 percent
    (E) 11 percent

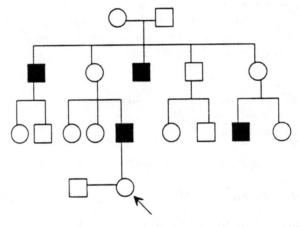

317. The proband's chance to carry an abnormal allele for vWD

318. The proband's risk to have an affected child

319. The proband's risk to have an affected child if she married her affected uncle

320. The proband's risk to have an affected child after she had three unaffected boys

321. The risk that the proband's aunt is a carrier given that she has three unaffected boys

Questions 322–325

The pedigree below shows individuals affected with autosomal recessive xeroderma pigmentosum. Match the individuals below with their risks to be carriers (heterozygotes).

(A) 100 percent
(B) 75 percent
(C) 67 percent
(D) 50 percent
(E) 33 percent

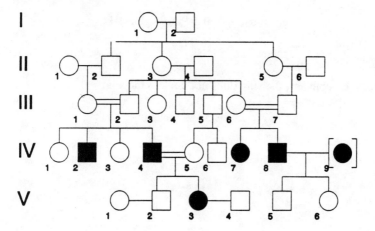

322. II-2

323. III-5

324. IV-1

325. V-5

Questions 326–327

Match the disorders below to the type of defect.

(A) Precursor accumulation
(B) Alternate pathway overproduction
(C) Product deficit
(D) None of the above

326. In galactokinase deficiency, galactitol in the lens causes cataracts

$$\text{galactose} \rightarrow \text{gal-1-PO}_4 \nrightarrow \text{gluc-1-PO}_4$$
$$\downarrow$$
$$\text{galactitol}$$

327. In albinism, the skin does not contain melanin

$$\text{tyrosine} \nrightarrow \text{melanin}$$

Clinical Genetics

Answers

284. The answer is E. *(Gelehrter, pp 270–275. Thompson, 5/e. pp 407–409.)* Screening programs for newborns are currently in place in all 50 states. Beginning in the 1960s these programs were established to identify infants with treatable genetic diseases. In deciding whether to establish a screening program and for which disease to screen, many issues must be addressed. Frequency and severity of the disorders must be established. In general, disorders included in screening programs should have an efficacious method of treatment that, if instituted early, will prevent serious consequences of the disorder. The test used to detect the disorder should be reliable, accurate, safe, and relatively inexpensive (such that the cost of screening is outweighed by the savings of early treatment and prevention of serious sequelae). There must be resources to assure effective screening and follow-up once affected individuals are identified. The availability of prenatal testing is not a prerequisite for a successful newborn screening program.

285. The answer is B. *(Thompson, 5/e. pp 392–394.)* A teratogen is an agent that can cause a malformation or raise the incidence of malformation. The dosage of a teratogen to which a fetus is exposed affects the response to that agent. The mother's ability to metabolize the agent, a factor related to her genotype, also plays a role. The genotype of the mother is also important in situations in which the mother has a genetic defect such as phenylketonuria (PKU). In maternal PKU, the elevated levels of phenylalanine are teratogenic to the fetus. Fetal genotype also affects response to a teratogen. The effects of a given teratogen may be different depending on the stage of development at which the fetus is exposed. Despite publicity about "agent orange," paternal exposure is unlikely to affect teratogenicity of a given agent.

286. The answer is C. *(Gelehrter, pp 27–47. Thompson, 5/e. pp 53–95.)* Although normal parents with two affected children would normally suggest autosomal recessive inheritance, the striking family history of lens dislocation (ectopia lentis) should caution one against this conclusion. If the paternal relatives have the same disorder, then X-linked inheritance is ruled out by the male-to-male transmission and made unlikely by the affected female. Mitochondrial inheritance would require that maternal relatives be affected. Autosomal dominant inheritance with incomplete penetrance in the father is most likely; variable expressivity implies the mildly affected person has some manifestations of the disease.

287–291. The answers are: 287-B, 288-C, 289-D, 290-E, 291-A. *(Thompson, 5/e. pp 391–393.)* Anomalous development of an organ primordium may be due to intrinsic factors or to extrinsic interference. Early and intrinsic abnormalities are termed *malformations,* and these point to Mendelian, chromosomal, or polygenic causation. Disruptions and deformations often reflect the operation of mechanical factors that are less likely to be genetic. If a single abnormality produces a chain of embryologically related events—as with early urethral obstruction that leads to abdominal distention with abdominal muscle wall weakness and excess skin (prune belly)—then the pattern of anomalies is called a *sequence.* If a single abnormality produces a pattern of defects that seem embryologically unrelated, this is called a *syndrome* (literally, *running together).* Syndromes have a high chance of having Mendelian or chromosomal causation, while sequences imply the polygenic or sporadic inheritance of single birth defects.

292–296. The answers are: 292-C, 293-A, 294-E, 295-D, 296-B. *(Gelehrter, pp 255–286.)* Genetic counseling is usually nondirective; that is, it provides facts without bias as to the patient's reproductive decisions. Preconceptional counseling is an important goal for all genetic or teratogenic diseases so that appropriate risks and management can be known before pregnancy is initiated. Prenatal counseling involves child-desiring or expectant couples who seek advice on risks, prenatal diagnostic options, and perinatal management. Supportive counseling is within the expertise of every physician and involves acknowledging a disease with genetic implications and providing a plan for evaluation and diagnosis. Since many genetic events are crises—for example, the birth of an abnormal child—there is a natural overlap of the family's inability to hear detailed counseling information and the physician's inability to provide it until tests or consultations are obtained. Informative counseling provides the details of disease and recurrence risk to concerned family members. It requires a precise diagnosis or diagnostic category.

297–301. The answers are: 297-A, 298-E, 299-B, 300-B, 301-C. (*Thompson, 5/e. pp 391–393.*) A malformation such as isolated cleft lip and palate is a morphologic defect of an organ or part of an organ that results from an intrinsically abnormal process of development. When there is extrinsic interference with an originally normal developmental process, such as in the case of amniotic bands, it is known as a *disruption*. A *syndrome* is a pattern of multiple malformations or anomalies thought to be pathogenetically related but embryologically unrelated. Syndromes may be caused by chromosomal abnormalities as in the case of trisomy 13. They may also be Mendelian disorders as in the case of the Stickler syndrome, an autosomal dominant defect in collagen that is partially described in question 300. A sequence, such as the Pierre Robin sequence described in question 297, is a pattern of anomalies that originates from a single prior defect or mechanical factor. These examples illustrate that a similar clinical picture may result from different etiologies.

302–306. The answers are: 302-B, 303-A, 304-B, 305-D, 306-B. (*Gelehrter, pp 27–47. Thompson, 5/e. pp 53–72.*) Each parent generates 1/2 A and 1/2 S gametes, implying probabilities of 1/4 AA, 1/2 AS, and 1/4 SS for genotypes in offspring. If q represents the S allele frequency of $0.1 = 1/10$, then p will be about 1, and $2pq$ about 1/5—the chance that an unrelated person will be genotype AS. The probability for their AS child to have an SS offspring is $(1/5)(1/4) = 1/20$. Mating of AS children would give a 1/4 probability of an SS grandchild, which illustrates the effect of consanguinity.

307–311. The answers are: 307-A, 308-D, 309-A, 310-E, 311-C. (*Thompson, 5/e. pp 53–72.*) The history suggests alkaptonuria (black urine), which was one of the first genetic disorders to be recognized as Mendelian by Sir Archibald Garrod. Consanguinity of the parents always suggests autosomal recessive disorders, as does the metabolic abnormality implied by urine with an unusual color or odor. The parents must both be genotypes Aa with a 1/4 risk for another aa child. The proband must be genotype aa and her prospective unrelated spouse has a $2pq$ $(2)(1/100) = 1/50$ chance to be an Aa carrier, which results in a $(1)(1/50)(1/2) = 1/100$ chance for an affected aa child. If she marries her uncle, then he has a 1/2 chance to be genotype Aa with a $(1)(1/2)(1/2) = 1/4$ chance for an affected child. The unaffected sibling will have a 2/3 chance to be Aa, with a $(2/3)(1/50)(1/4) = 1/300$ chance for an affected child if her spouse is unrelated and a $(2/3)(1/2)(1/4) = 1/12$ chance if she marries her uncle.

312–316. The answers are: 312-D, 313-E, 314-A, 315-C, 316-B.
(Gelehrter, pp 125–158. Thompson, 5/e. pp 271–316.) Inherited disorders of metabolism usually result from severe enzyme deficiency (0 to 5 percent residual activity) with consequent elevation or depletion of metabolites. Some elevated metabolites, such as phenylalanine, are probably toxic to the brain and cause mental retardation, while others, like homogentisic acid, are deposited in cartilage and cause arthritis. Metabolite deficiency is exemplified by the lack of purines needed for immune response in adenosine deaminase deficiency. Therapy for metabolic disorders is based on removing the accumulating substance, supplementing substrate deficiencies, or both. Because enzymes are present in amounts greater than needed for usual pathway demands (enzyme reserve), reduction to 50 percent levels usually has no effect. This is the usual reason that heterozygotes (carriers) for autosomal recessive diseases are normal.

317–321. The answers are: 317-A, 318-D, 319-C, 320-D, 321-E.
(Gelehrter, pp 27–47. Thompson, 5/e. pp 53–73.) The pedigree is much more suggestive of X-linked recessive than autosomal dominant inheritance because of its oblique pattern and predominance of affected males. Reference to McKusick's *Mendelian Inheritance in Man* reveals a rare X-linked recessive form of vWD that supports this conclusion. The proband is an obligate carrier since her father is affected, and she has a 1/2 chance to transmit the abnormal allele and a 1/2 chance to have a son. Her risk of having an affected child is thus $(1/2)(1/2) = 1/4$. Her uncle would also have an abnormal allele on his X, which would result in risks of 1/2 for affected males and 1/2 for an affected female with abnormal alleles on both X chromosomes. Subsequent unaffected males would not change the proband's status of obligate carrier, but three normal boys would change her aunt's risk from 1/2 chance of being a carrier to 1/9 (11 percent) using Bayesian counseling.

322–325. The answers are: 322-A, 323-A, 324-C, 325-A. *(Gelehrter, pp 27–47. Thompson, 5/e. pp 53–72.)* Xeroderma pigmentosum (XP) is an autosomal recessive disorder in which defective DNA repair causes sensitivity to sunlight, skin lesions, and cancer. Autosomal recessive inheritance is supported by the consanguinity in the pedigree. The spouse (IV-9) of individual IV-8 has more severe disease and is likely to be unrelated although she is an adoptee. If she had the same recessive disease, both parents would be homozygous abnormal and all their children would be affected. The most likely explanation is genetic heterogeneity, with the parents' having mutations at different loci that produce similar clinical symptoms. The presence of up to five different loci responsible for XP is suggested by coculture of fibroblasts from different patients and noting whether they can restore each other's repair defects (complementation). If different loci are involved, then the offspring would be carriers for both abnormal alleles but would not be affected.

In order for IV-2 to have been affected (homozygous abnormal), II-2 must have transmitted the abnormal allele to III-1 and thus must be a carrier. The same is true for III-5 in order for V-3 to be affected. V-5 must be a carrier because a parent is homozygous affected. IV-1 is an unaffected sibling of IV-2 and cannot be homozygous abnormal. Given the abnormal allele x and the normal allele X, her remaining alternatives for genotypes are Xx (x from the mother), xX (x from the father), or XX—a 2/3 chance of being a carrier.

326–327. The answers are: 326-B, 327-C. *(Gelehrter, pp 125–158. Thompson, 5/e. pp 271–316.)* The pathogenesis of different inborn errors of metabolism may be variable. In galactokinase deficiency, nuclear cataracts seen either before or shortly after birth are the only clinical manifestation. Because the conversion of galactose to galactose-1-phosphate is blocked, galactitol, the product of an alternative pathway, accumulates in the lens and causes cataracts. In tyrosine-negative albinism, tyrosine cannot be converted to melanin, a pigment that gives color to hair, skin, and eyes. Because of this product deficit, patients with albinism have reduced visual acuity and nystagmus, dermatosis sensitive to ultraviolet light, and photophobia.

Reproductive Genetics

DIRECTIONS: Each question below contains five suggested responses. Select the **one best** response to each question.

328. Indications for amniocentesis may include all the following EXCEPT

(A) prior child with mental retardation
(B) maternal age > 35
(C) prior child with spina bifida
(D) prior child with inborn error of metabolism
(E) prior child with chromosomal abnormality

329. Which of the following should be involved in every prenatal diagnosis?

(A) Level I ultrasound
(B) Chorionic villus sampling (CVS)
(C) Doppler analysis
(D) Amniocentesis
(E) Genetic counseling

330. Prenatal diagnosis results in what frequency of termination of pregnancy?

(A) > 90 percent
(B) 50 percent
(C) 25 percent
(D) 10 percent
(E) < 5 percent

331. Genetic counseling helps a couple do all the following EXCEPT

(A) understand the medical facts
(B) understand the mode of inheritance and recurrence risks
(C) adjust to the condition
(D) rule out the need for prenatal diagnosis when abortion is unacceptable
(E) select a course of action

332. All the following are required for biochemical prenatal diagnosis of inborn errors of metabolism EXCEPT

(A) a biochemical assay must be available that can be used on cell culture or amniotic fluid
(B) the biochemical defect must be expressed in amniotic fluid, amniocytes, or chorionic villi
(C) the assay must separate individuals with 50 percent enzyme levels from those with 0 to 10 percent
(D) the defect must be autosomal recessive
(E) results must be available within 2 to 4 weeks

DIRECTIONS: Each group of questions below consists of lettered headings followed by a set of numbered items. For each numbered item select the **one** lettered heading with which it is **most** closely associated. Each lettered heading may be used **once, more than once, or not at all.**

Questions 333–337

Match the terms below with their characteristics.

(A) 0.5 to 1.0 mL fetal blood
(B) 9 to 12 weeks gestation
(C) 20 to 30 mL fluid volume
(D) All trimesters
(E) Maternal blood

333. Chorionic villus sampling (CVS)

334. Amniocentesis

335. Level I ultrasound

336. Percutaneous umbilical blood sampling (PUBS)

337. Maternal serum alpha-feto-protein (MSAFP)

Questions 338–342

Match the terms below with their characteristics.

(A) Results within 24 to 48 h
(B) Results within 1 to 2 weeks
(C) Fetal blood flow
(D) Dating of pregnancy
(E) Fetal anomalies

338. Level I ultrasound

339. Level II ultrasound

340. Doppler analysis

341. Amniocentesis

342. Cordocentesis

Questions 343–347

A 37-year-old woman and her 45-year-old husband request genetic counseling regarding their second pregnancy. She undergoes amniocentesis and level I ultrasound study at an estimated 17 weeks of gestation. A single amniotic sample was obtained, and a photograph from the ultrasound is shown in the figure below. Match the interpretations with the results below.

(A) Amniotic AFP concentration of 30 µg/mL; karyotype 46,XY
(B) Amniotic AFP 2 MOM (multiples of median) for gestational age of 15 weeks estimated from ultrasound; karyotype 46,XY
(C) Amniotic AFP 6 MOM for gestational age of 15 weeks; karyotype 46,XY
(D) Amniotic AFP 2 MOM for gestational age of 15 weeks; karyotype 47,XY+21
(E) None of above

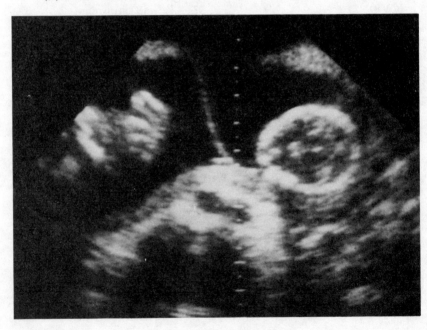

343. Normal twin gestation

344. Normal karyotype, AFP level uninterpretable unless correlated with population mean and gestational age

345. Down syndrome in one fetus, expected low amniotic fluid AFP

346. Possible spina bifida with normal karyotype in one of twins

347. Normal amniotic fluid AFP and karyotype in one of twins

Questions 348–351

A 33-year-old woman seeks genetic counseling regarding her first pregnancy. The couple's family and past medical histories were not remarkable. Measurement of MSAFP concentration and level I ultrasound were performed at an estimated gestational age of 16 weeks (according to last menstrual period). Match the results below with the probable explanations.

    (A) 20 ng/mL
    (B) 5.5 MOM for gestational age by ultrasound of 15 1/2 weeks
    (C) 1.0 MOM for gestational age by ultrasound of 15 1/2 weeks
    (D) 0.5 MOM for gestational age by ultrasound of 15 1/2 weeks

348. Possible fetal spina bifida

349. Uninterpretable result

350. Possible fetal Down syndrome

351. Possible fetal death or twin pregnancy

Questions 352–361

Match the following procedures with their pregnancy risks.

(A) 2 to 3% risk for fetal loss
(B) 0.5 to 1 percent risk for
    fetal loss above sponta-
    neous abortion rate
(C) Same risk as B, plus ill-
    defined risk for limb de-
    fects
(D) No known risk
(E) Risk undetermined

352. Level I ultrasound

353. Level II ultrasound

354. Doppler analysis

355. Early amniocentesis (13 to
14 weeks of gestation)

356. Amniocentesis (15 to 17
weeks of gestation)

357. Cordocentesis

358. Chorionic villus biopsy (<9
weeks of gestation)

359. Chorionic villus biopsy (9
to 12 weeks of gestation)

360. Preimplantation diagnosis

361. Prenatal diagnosis without
follow-up counseling

Questions 362–366

Match the conditions below
with a feasible method of prenatal
diagnosis.

(A) Fetal DNA testing after
    parental mutations are
    determined
(B) Fetal karyotyping to de-
    termine sex
(C) Fetal karyotyping to de-
    termine chromosome
    number
(D) Hexaminidase assay of
    chorionic villi
(E) Type II ultrasound and
    measurement of alpha-
    fetoprotein in amniotic
    fluid

362. Tay-Sachs disease

363. Down syndrome

364. Spina bifida

365. Cystic fibrosis

366. Nonspecific X-linked mental retardation

Reproductive Genetics
Answers

328. The answer is A. *(Gelehrter, pp 275–283. Thompson, 5/e. pp 411–425.)* Amniocentesis has a risk for inducing miscarriage of less than 0.5 percent, so diagnosable conditions with a recurrence risk of about 1 percent or higher are acceptable indications for the procedure. Mental retardation is not a specific diagnosis, so prenatal detection is not feasible. At age 35, the risk for Down syndrome plus other chromosomal anomalies approaches 1 percent—a figure that applies to any couple having a prior child with a chromosomal anomaly regardless of their age. Parents of a child with a neural tube defect (spina bifida, anencephaly) face a 2 percent recurrence risk, and the usual autosomal recessive inheritance of metabolic disorders predicts a 25 percent risk after an affected child.

329. The answer is E. *(Gelehrter, pp 275–283. Thompson, 5/e. pp 411–425.)* Genetic counseling is an essential component of every prenatal diagnostic test. Couples must understand their risks and options before selecting a prenatal diagnostic procedure, and there must be adequate provisions for explaining the results to them. Since additional obstetric procedures such as termination of pregnancy may follow prenatal diagnosis, obstetricians are key elements in the genetic counseling process.

330. The answer is E. *(Gelehrter, pp 275–283. Thompson, 5/e. pp 411–425.)* Although autosomal and X-linked recessive disorders imply a 25 percent recurrence risk, the majority of prenatal diagnostic procedures are performed because of advanced maternal age (>35). With a risk ranging from about 1 to 2 percent at age 35 to a maximum of 5 percent at age 45, the overall frequency of elective pregnancy termination following prenatal diagnosis is only about 3 percent. Thus 97 percent of couples are reassured by prenatal diagnostic procedures.

331. The answer is D. *(Gelehrter, pp 275–283. Thompson, 5/e. pp 411–425.)* Genetic counseling addresses the medical circumstances, inheritance and recurrence risks, impact of a given risk for a particular family (burden), options for management, and adjustment to the circumstances of risk and disease. Such counseling is traditionally nondirective, which means that the counselor does not attempt to influence the family's decisions. While abortion is unacceptable to many couples, prenatal diagnosis may still be useful in guiding perinatal management and in preparing the couple.

332. The answer is D. *(Gelehrter, pp 275–283. Thompson, 5/e. pp 411–425.)* Most inborn errors of metabolism have been explained by deficiencies of particular enzymes that can be assayed in skin fibroblasts. Such deficiencies have usually been found in the corresponding amniocytes or chorionic villi. Occasionally, the elevated substrate or deficient product of this enzyme reaction can be reliably assayed in body fluids, including amniotic fluid. Biochemical disorders such as Marfan syndrome are autosomal dominant and involve defects in connective tissue molecules. Because amniocyte or chorionic villus culture often requires 2 to 3 weeks, a reliable and rapid biochemical assay is needed for prenatal diagnosis. Most centers will not perform termination of pregnancy after 24 weeks of gestation.

333–337. The answers are: 333-B, 334-C, 335-D, 336-A, 337-E. *(Thompson, 5/e. pp 411–425.)* Chorionic villus sampling (CVS) involves aspiration of placental trophoblast tissue at 9 to 12 weeks of pregnancy. Since trophoblast cells are of fetal origin, direct or culture analysis for chromosomes can document the fetal karyotype. Amniocentesis is routinely performed at 15 to 17 weeks of pregnancy and withdraws 20 to 30 mL of amniotic fluid and cells for fetal analysis. Early amniocentesis that withdraws smaller amounts of fluid may be performed at 12 to 14 weeks of pregnancy. Level I ultrasound allows routine monitoring of embryonic and fetal growth when specific fetal anomalies are not anticipated; level II ultrasound for detection of anomalies is performed in the second trimester. Percutaneous umbilical blood sampling (PUBS), or cordocentesis, is used to aspirate 0.5 to 1 mL of fetal blood from the umbilical vein after maternal transabdominal puncture. Maternal serum alpha-fetoprotein (MSAFP) levels are related to numerous gestational factors including the presence of fetal neural tube defects and trisomies.

338–342. The answers are: 338-D, 339-E, 340-C, 341-B, 342-A. *(Gelehrter, pp 275–283. Thompson, 5/e. pp 411–425.)* Level I ultrasound allows such general determinations as dating of pregnancy, fetal and placental placement, and detection of twins. Level II ultrasound is more time-consuming and involves careful imaging of specific fetal organs for detection of anomalies. Doppler analysis uses continuous ultrasound waves to determine the direction and velocity of blood flow in the fetoplacental unit. Since cordocentesis provides dividing fetal lymphocytes, karyotype results can be obtained in 24 to 48 h. Amniocentesis yields fibroblast-like fetal amniocytes that require 1 to 2 weeks of culture before dividing cells can be arrested in metaphase for karyotyping.

343–347. The answers are: 343-E, 344-A, 345-E, 346-C, 347-B. *(Gelehrter, pp 275–283. Thompson, 5/e. pp 411–425.)* The figure clearly outlines two amniotic cavities indicative of a twin gestation. The single amniotic sample is thus representative of only one twin. Because two amnions can occur with either monozygotic or dizygotic twins, zygosity cannot be determined from the ultrasound. Amniotic fluid AFP values must be correlated with population mean and gestational age before they are meaningful; large elevations occur with neural tube defects (anencephaly, spina bifida) and other fetal anomalies that allow leakage of fetal fluids directly into the amniotic fluid. Since spina bifida is a polygenic defect, it is not necessarily concordant (present in both twins) even if the twins are monozygotic. Sampling of the other amniotic sac would be required before the karyotype and neural tube status of the other twin could be assessed.

348–351. The answers are: 348-B, 349-A, 350-D, 351-B. *(Gelehrter, pp 275–283. Thompson, 5/e. pp 411–425.)* Concentrations of maternal serum alpha-fetoprotein (MSAFP) are meaningful only if correlated with population means and gestational age. Alpha-fetoprotein is a fetal homologue of albumin that rises early and decreases later in gestation. Leakage of fetal blood or amniotic fluid into the maternal bloodstream produces small but detectable amounts of AFP in maternal serum. The decreased rate of fetal growth and development seen with autosomal trisomies decreases MSAFP values, while leakage of AFP from open neural tube defects will increase MSAFP values. Other fetal conditions that produce elevated amniotic fluid AFP and MSAFP values include twins, fetal death, and fetal anomalies that disrupt skin or kidney (omphalocele, nephrosis, epidermolysis).

352–361. The answers are: 352-D, 353-D, 354-D, 355-B, 356-B, 357-A, 358-C, 359-B, 360-E, 361-E. *(Gelehrter, pp 275–283. Thompson, 5/e. pp 411–425.)* All medical procedures must be evaluated in terms of risk-benefit and cost-benefit ratios. Prenatal detection of Down syndrome has been justified based on saving of dependent care costs, but more optimistic attitudes regarding the potential of children with Down syndrome complicate such comparisons. For amniocentesis and chorionic villus sampling, slightly increased risks for fetal loss above spontaneous abortion rates have been measured in numerous studies. Early CVS may add a slight risk for fetal limb defects. Fetal blood sampling is the most risky prenatal procedure, although the risk of in vitro fertilization, blastomere sampling for diagnosis by polymerase chain reaction (PCR), and implantation (pre-implantation diagnosis) is not known. Pre- and postdiagnostic counseling is essential for good prenatal management, and its absence results in an unknown frequency of inappropriate elective abortions as well as other problems.

362–366. The answers are: 362-D, 363-C, 364-E, 365-A, 366-B. *(Gelehrter, pp 275–283. Thompson, 5/e. pp 411–425.)* Different prenatal diagnosis strategies are available depending upon the disorder of interest. Biochemical disorders such as Tay-Sachs disease may be diagnosed by assay of amniocytes or chorionic villi. Recent progress in molecular characterization of Tay-Sachs mutations makes DNA analysis or hexaminidase assay feasible. Only DNA analysis is available for cystic fibrosis since the sweat test used for children is not applicable to fetal tissues. Because many different mutations have been described for cystic fibrosis, those present in the parents must be defined for accurate prenatal diagnosis. Although a fetal karyotype is diagnostic for Down syndrome, in ill-defined X-linked mental retardation it will only discriminate between females, who may be either mildly affected or unaffected, and males, who may be either severely affected or unaffected.

Molecular Genetics

DIRECTIONS: Each question below contains four or five suggested responses. Select the **one best** response to each question.

367. All the following are true in regard to in situ hybridization EXCEPT

(A) DNA probes are annealed to metaphase chromosomes
(B) the method is useful for both single-copy and repetitive DNA sequences
(C) chromosomes are spread on a slide and denatured in place
(D) probes may be labeled with fluorescent material
(E) exact localization of genes is possible

368. All the following statements are true regarding the Human Genome Project EXCEPT

(A) it represents a national effort
(B) it involves constructing physical and genetic linkage maps
(C) it includes plans to sequence all genes
(D) maps are maintained in a computer data base available to all interested parties

369. The most common type of mutation found in DNA is

(A) insertion
(B) gene deletion
(C) small intragenic deletion
(D) point mutation

370. Single-base substitutions in human DNA

(A) usually result in disease
(B) are rarely seen in intronic sequences
(C) usually result in the creation of a new restriction fragment length polymorphism (RFLP)
(D) may affect transcription
(E) may result in a single codon deletion

371. Repetitive sequences in DNA include all the following EXCEPT

(A) L1 family (LINES)
(B) Alu sequences
(C) alphoid sequences
(D) fragile sites
(E) satellite DNA

372. The figure below illustrates a family in which individual I-1 has an autosomal dominant disease. Crossing-over is most likely to have occurred in which of her offspring?

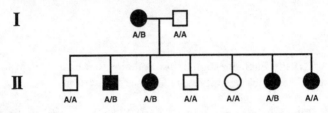

(A) Individual II-1
(B) Individual II-2
(C) Individual II-4
(D) Individual II-6
(E) Individual II-7

373. In the family depicted in the figure below, the genotype of the deceased individual, I-1, is most likely to be which of the following?

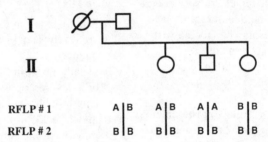

(A) B | B
    B | B
(B) A | B
    B | B
(C) A | A
    B | B
(D) Undeterminable from data given

Questions 374–375

Southern blot analysis is performed on DNA from the family depicted in the figure below.

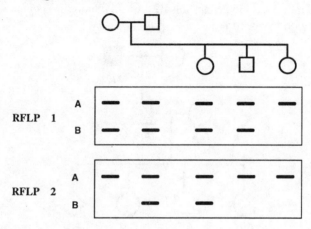

374. The genotype of individual II-2 is:

(A) A | B
　　A | B

(B) A | A
　　B | B

(C) B | B
　　A | A

(D) A | B
　　A | A

(E) undeterminable from the data given

375. The genotype of the father, individual I-2, is

(A) A | B
　　A | B

(B) A | A
　　B | B

(C) B | B
　　A | A

(D) A | B
　　A | A

(E) undeterminable from the data given

Questions 376–378

Prenatal diagnosis is performed for an autosomal dominant condition with onset in adulthood on the family depicted in the figure below. All individuals in generation III are still children. Assume no recombination has occurred.

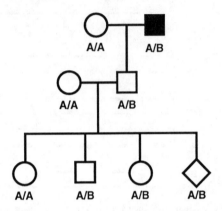

376. The fetus (III-4) is

(A) affected
(B) unaffected
(C) at 50 percent risk
(D) at 25 percent risk
(E) of undeterminable status
from data given

377. Individual III-1 is

(A) affected
(B) unaffected
(C) at 50 percent risk
(D) at 25 percent risk
(E) of undeterminable status
from data given

378. Individual II-2 has begun to show signs and symptoms of the disorder. With this information in mind, choose the status of the fetus.

(A) Affected
(B) Unaffected
(C) At 50 percent risk
(D) At 25 percent risk
(E) Undeterminable from data given

Questions 379–380

A couple comes to your office after having had a child with cystic fibrosis. The wife is now pregnant and the couple desires prenatal diagnosis. Amniocentesis is performed at 17 weeks of pregnancy and the results are shown in the figure below. Assume that no recombination has occurred.

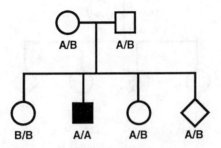

379. What is the status of the fetus (II-4)?

(A) Affected
(B) Unaffected, carrier
(C) Unaffected, noncarrier
(D) Undeterminable from the data given

380. What is the status of individual II-1?

(A) Affected
(B) Unaffected, carrier
(C) Unaffected, noncarrier
(D) Undeterminable from the data given

Questions 381–382

Prenatal diagnosis is performed for an autosomal recessive condition. Results are shown in the figure below. Assume that no recombination has occurred.

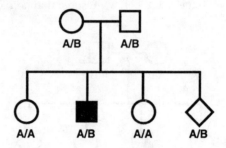

381. What is the status of the fetus (II-4)?

(A) Affected
(B) Unaffected, carrier
(C) Unaffected, noncarrier
(D) Undeterminable from the data given

382. What is the status of individual II-1?

(A) Affected
(B) Unaffected, carrier
(C) Unaffected, noncarrier
(D) Undeterminable from the data given

383. If allele-specific oligonucleotides R′ and r′ were synthesized such that they were complementary to alleles R and r, choose the diagram that would correspond to the results of dot-blot hybridization of DNA from the indicated genotypes under stringent conditions. Possible results (A–E) are depicted in the figure below.

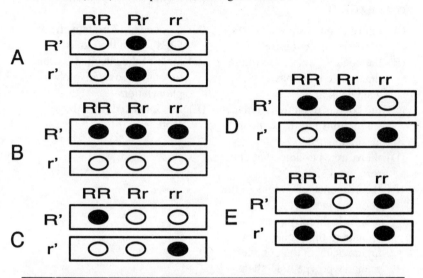

384. Sickle cell anemia is caused by a point mutation in the hemoglobin gene that results in the substitution of a single amino acid in the mature protein. This mutation could be detected by all the following methods EXCEPT

(A) allele-specific oligonucleotide (ASO) hybridization
(B) Southern blot analysis
(C) Western blot analysis
(D) polymerase chain reaction (PCR) with restriction enzyme digestion
(E) DNA sequencing

385. In Huntington's disease, patients tend to have an earlier age of onset of symptoms if the gene is inherited from an affected father as opposed to an affected mother. The most likely mechanism for this finding is

(A) mitochondrial inheritance
(B) imprinting
(C) germline mosaicism
(D) uniparental disomy
(E) variable expressivity

386. The "genetic code" stores information that determines the amino acid sequence of proteins. All of the following statements are true regarding the genetic code EXCEPT

(A) information is stored as sets of three adjacent bases
(B) the code is degenerate (more than one codon may exist for a single amino acid)
(C) the code is essentially universal (codons are the same in all organisms)
(D) there are 64 codons that code for amino acids
(E) the sequence of codons in the coding regions of a DNA molecule directly corresponds to the sequence of amino acids in the complementary polypeptide chain

387. In Leber's hereditary optic neuropathy, all individuals are related through the maternal line. Affected males cannot pass on the disease. The most likely mechanism for this finding is

(A) mitochondrial inheritance
(B) imprinting
(C) germline mosaicism
(D) uniparental disomy
(E) variable expressivity

388. All the following statements are true concerning mitochondrial DNA EXCEPT

(A) it is double-stranded
(B) it encodes its own set of tRNAs
(C) it encodes 13 proteins translated within the nucleus
(D) most of the encoded proteins function in oxidative phosphorylation
(E) it is inherited from the mother

DIRECTIONS: Each group of questions below consists of lettered headings followed by a set of numbered items. For each numbered item select the **one** lettered heading with which it is **most** closely associated. Each lettered heading may be used **once, more than once, or not at all.**

Questions 389–392

Match the terms below with their partial definitions.

(A) Number of DNA or RNA base pairs estimated from nucleic acid hybridization data
(B) Labeled DNA or RNA fragment used for hybridization
(C) Oligonucleotide designed to detect a particular allele
(D) Oligonucleotide designed for DNA sequencing or PCR reaction
(E) Hybridization conditions that determine the amount of mismatching allowed between hybrids

389. Primer

390. Probe

391. Stringency

392. Complexity

Questions 393–396

Match the terms below with their partial definitions.

(A) Thermal cycler
(B) One percent recombination
(C) Marker for linkage
(D) Odds for linkage
(E) DNA fingerprinting

393. Restriction fragment length polymorphism (RFLP)

394. Variable numbers of tandem repeats (VNTRs)

395. Polymerase chain reaction (PCR)

396. 1 centimorgan (cM)

Questions 397–401

Match the terms below with their approximate DNA contents.

(A) 3 bp
(B) 40,000 bp (40 kb)
(C) 2×10^6 bp
(D) 1.5×10^8 bp
(E) 3×10^9 bp

397. Average chromosome band from high-resolution karyotype

398. Human haploid genome

399. Codon

400. Average chromosome

401. Average-sized gene

Questions 402–404

Match the terms below with their partial definitions.

(A) First registered chromosome 13 multicopy DNA probe
(B) Thirteenth registered chromosome 5 single copy DNA probe
(C) Thirteenth registered X chromosome single copy DNA probe
(D) Fourteenth registered X chromosome single copy DNA probe
(E) Fifth registered chromosome 13 single copy DNA probe

402. DXS14

403. D13S5

404. D13Z1

Questions 405–409

Match the descriptions below with the appropriate ribonucleic acid.

(A) mRNA
(B) DNA
(C) tRNA
(D) rRNA

405. 3'ACCTG5'

406. Contained in ribosomes

407. Approximately 80 nucleotides long

408. Forms a template for the synthesis of polypeptide chains

409. Double-stranded structure

Questions 410–413

Match the terms below with their partial definitions.

(A) Conversion of triplet codons into amino acids
(B) Duplication of DNA
(C) Pairing of complementary strands of nucleic acids
(D) Conversion of DNA to RNA
(E) None of the above

410. Replication

411. Translation

412. Hybridization

413. Transcription

Questions 414–418

The hypothetical "stimulin" gene contains two exons that encode a protein of 100 amino acids. They are separated by an intron of 100 bp beginning after the codon for amino acid 10. Stimulation mRNA has 5' and 3' untranslated regions of 70 and 30 nucleotides, respectively. Match the characteristics of the stimulin gene below with the appropriate measures.

(A) 500 bp
(B) 400 bp
(C) 300 bp
(D) 100 bp
(E) 70 bp

414. Size of stimulin mRNA precursor—minus poly(A) tail

415. Distance between transcription and translation start sites

416. Size of mature stimulin RNA—minus residual poly(A)

417. Minimum size of stimulin gene

418. Distance of first splice site from transcription start site

Questions 419–423

Match the statements below with the type of information they convey.

(A) Genetic mapping
(B) Genetic distance
(C) Physical mapping
(D) Physical distance
(E) None of the above

419. The locus for nail-patella syndrome exhibits linkage to the ABO blood group locus

420. The locus for Stickler syndrome is 5 cM away from the type II collagen gene

421. The gene determining Stickler syndrome and that for type II collagen were found on the same yeast artificial chromosome

422. In situ hybridization reveals that the type II collagen gene is at chromosome band 12q13

423. A gene of the Wnt family is located 200 kb upstream from the cystic fibrosis gene on chromosome 7

Questions 424–427

Match the terms below with their partial definitions.

(A) Rapidly amplifies fragments of DNA
(B) Detects alterations in gene expression
(C) Identifies restriction fragments of specific genes
(D) Separates DNA fragments by size

424. Southern blot analysis

425. Northern blot analysis

426. Polymerase chain reaction

427. Agarose gel electrophoresis

Questions 428–431

Match the terms below with the partial definitions.

(A) Pyrimidine
(B) Purine
(C) Sugar
(D) Nucleotide

428. Deoxyribose

429. Adenine

430. Ribose

431. Thymine

Questions 432–436

Match the descriptions below with the correct term.

(A) Intron
(B) Exon
(C) CAT box
(D) AATAAA
(E) Pseudogene

432. Coding region of DNA

433. DNA regulatory element

434. Codes for poly(A) tail of RNA

435. Located in downstream flanking region of gene

436. Removed in processing of precursor RNA

Questions 437–441

Match the descriptions below with the appropriate term.

(A) Histone
(B) Chromatin
(C) Heterochromatin
(D) Transposon
(E) Centromere

437. Ability to "jump" to a new site in DNA

438. Proteins, rich in basic amino acids, which are packaged with DNA in chromosomes

439. Condensed packages of proteins and DNA

440. Primary constriction

441. Point of attachment for sister chromatids

Questions 442–446

The DNA sequence M shown below is the sense strand from a coding region known to be a mutational "hotspot" for a gene. It encodes amino acids 21 through 25. Given the genetic and amino acid codes CCC = proline (P), GCC = alanine (A), TTC = phenylalanine (F), and TAG = stop codon, match the mutant sequences with the notations below.

M   5'-CCC-CCT-AGG-TTC-AGG-3'

(A) $P_{21}A$, point mutation
(B) Frameshift mutation, chain terminating
(C) ΔF_{24}, trinucleotide deletion
(D) Frameshift mutation, nucleotide deletion
(E) $C_{63}A$, silent mutation

442. -CCA-CCT-AGG-TTC-AGG-

443. -GCC-CCT-AGG-TTC-AGG-

444. -CCA-CCC-TAG-GTT-CAG-

445. -CCC-CTA-GGT-TCA-GG--

446. -CCC-CCT-AGG-AGG-----

Questions 447–450

The ABO blood group locus is linked to the gene for the autosomal dominant nail-patella syndrome (NPS) with a recombination frequency of about 10 percent. A type O father has NPS as does his type B daughter. Her mother is type AB and unaffected, and her husband is normal. If the daughter and her husband wish to consider prenatal diagnosis for NPS by determining the fetal ABO blood type, match the following circumstances with the fetal risk to have NPS.

(A) 100 percent risk
(B) 90 percent risk
(C) 50 percent risk
(D) 10 percent risk
(E) Virtually 0 risk

447. Husband type O, fetus type B

448. Husband type A, fetus type O

449. Husband type A, fetus type AB

450. Husband type B, fetus type B

Questions 451–457

A hypothetical GUNNER gene has been cloned that is responsible for a recessive disease in people with rr genotypes. Shown below is a map for the GUNNER gene with 4 *Eco*RI restriction endonuclease sites (E1 through E4) and the distances between them in kilobases. The nucleotide sequences of R and r alleles are shown beneath the gene; recognition site E2 for *Eco*RI endonuclease is underlined and the GUNNER point mutation is marked by a star. Match the sizes of DNA fragments (in kb) that would be visualized after digestion of the patient's DNA with *Eco*RI and Southern blotting given the conditions below. (The probes were cloned from a normal patient.)

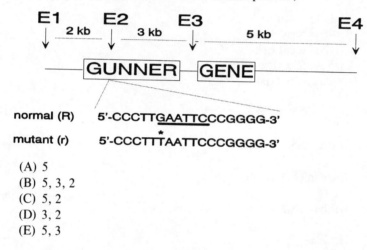

| normal (R) | 5'-CCCTT<u>GAATTC</u>CCGGGG-3' |
| mutant (r) | 5'-CCCTT*TAATTCCCGGGG-3' |

(A) 5
(B) 5, 3, 2
(C) 5, 2
(D) 3, 2
(E) 5, 3

451. E1–E3 DNA segment as probe, RR individuals

452. E1–E3 DNA segment as probe, Rr individuals

453. E1–E2 DNA segment as probe, rr individuals

454. E1–E2 DNA segment as probe, Rr individuals

455. E2–E3 DNA segment as probe, Rr individuals

456. E3–E4 DNA segment as probe, RR individuals

457. E3–E4 DNA segment as probe, rr individuals

Questions 458–462

Polymorphic DNA markers are being isolated from the region of chromosome 9 that contains the linked ABO and nail-patella syndrome (NPS) loci. One such marker is linked to the NPS locus with a recombination frequency of 0.05 and yields a 5-kb fragment when the restriction site is absent and 3- and 2-kb fragments when it is present. A father with NPS has a genotype of 3,2; his affected daughter a genotype of 5,3,2; her unaffected mother a genotype of 5,5. If the daughter and her unaffected husband wish to consider prenatal diagnosis for NPS by determining fetal marker status, match the genotypes below with the fetal risk for NPS.

(A) 100 percent risk
(B) 95 percent risk
(C) 50 percent risk
(D) 5 percent risk
(E) Virtually 0 risk

458. Husband 5,3,2; fetus 5,5

459. Husband 5,3,2; fetus 5,3,2

460. Husband 5,3,2; fetus 3,2

461. Husband 3,2; fetus 5,3,2

462. Husband 5,5; fetus 3,2

Molecular Genetics
Answers

367. The answer is E. *(Gelehrter, pp 195–196. Thompson, 5/e. p 175.)* In situ hybridization is a technique used for gene mapping. Metaphase chromosomes are denatured in place (in situ) on a slide. DNA probes, which may be labeled with radioactive or fluorescent material, are then annealed to the DNA. This method can be used to map genes with a resolution of 1 to 2 million base pairs. However, since this is still substantially larger than most genes, exact localization is not possible.

368. The answer is A. *(Gelehrter, pp 289–292. Thompson, 5/e. p 186.)* The Human Genome Project is an international effort to map and sequence all the genes in the human genome. Physical and genetic linkage maps are being constructed of all 23 pairs of human chromosomes. In the United States, the project was initially headed by James Watson, who, along with Francis Crick, originally described the structure of the DNA double helix. All current mapping data are maintained on-line on a continually edited computer data base.

369. The answer is D. *(Gelehrter, pp 21–23. Thompson, 5/e. pp 115–140.)* A mutation is a permanent change in DNA. Multiple types of mutations can occur. The most common type of mutation is a point mutation in which a single nucleotide is replaced. Point mutations may be silent and therefore not affect the final polypeptide product. They may also be missense mutations, which cause a single amino acid change; nonsense mutations, which cause a premature "stop" in DNA transcription; mutations that alter RNA splicing; or mutations that alter the regulation of transcription.

370. The answer is D. *(Gelehrter, pp 21–23. Thompson, 5/e. pp 115–141.)* Most DNA variation is normal (DNA polymorphisms). A very small percentage of changes cause disease. Variations in DNA occur approximately every 100 to 200 nucleotides in introns and flanking DNA. Frequently these changes cannot be detected by restriction enzymes and therefore do not result in the formation of a new RFLP (restriction fragment length polymorphism). Point mutations do not delete

base pair codons but may affect their transcription by altering promoter/
enhancer sites.

371. The answer is D. *(Gelehrter, p 71. Thompson, 5/e. pp 38–39.)*
Approximately 10 percent of the genome consists of highly repetitive,
clustered sequences of DNA organized in tandem arrays. These regions
of DNA are collectively called *satellite DNAs*, include alphoid sequences, and tend to be located in specific regions such as centromeres
and telomeres. Approximately 15 percent of the genome consists of
other repetitive sequences of DNA, which are scattered throughout the
regions of single-copy DNA. These repetitive DNA families include the
Alu family and the L1 family (or LINES).

372. The answer is E. *(Gelehrter, pp 202–207. Thompson, 5/e. pp 178–
190.)* In this family, affected individuals appear to have inherited the
paternal B allele, whereas unaffected individuals have inherited the paternal A allele. The sole exception to this is individual II-7, who inherited both the disease gene and the paternal A allele. If the polymorphism
studied is linked to the disease gene, individual II-7 must be a recombinant in whom crossing-over has occurred between the polymorphism
and the disease gene. In order to definitively prove linkage, more families would need to be studied.

373. The answer is B. *(Gelehrter, pp 202–207. Thompson, 5/e. pp 178–
190.)* In this example, phase, or those alleles that are traveling together,
has already been determined for you. Individual II-2 has inherited two
chromosomes, each of which contains the AB haplotype. Therefore,
each parent must have one chromosome that contains the AB haplotype. Similarly, individual II-3 has the BB haplotype on both chromosomes, which implies that each parent must have one chromosome that
carries the BB haplotype. One can now deduce that the genotype of
individual I-1 was AB/BB.

374. The answer is D. *(Gelehrter, pp 202–207. Thompson, 5/e. pp 178–
190.)* In this example, phase (those alleles that travel together) must be
determined. Individual II-2 is homozygous for allele A at RFLP 2.
Therefore, his genotype must be AA/BA.

375. The answer is A. *(Gelehrter, pp 202–207. Thompson, 5/e. pp 178–
190.)* In this example, phase must be determined. The mother in the
family, individual I-1, is homozygous for the A allele at RFLP 2. There-

fore, her genotype is AA/BA. Using the same logic, genotypes for individual II-2 (AA/BA) and individual II-3 (AA/AA) may be determined. Since individual II-3 must have inherited one chromosome from each parent, the father must have one chromosome that contains the haplotype AA. His other chromosome must therefore contain the haplotype BB.

376. The answer is C. *(Gelehrter, pp 202–207. Thompson, 5/e. pp 178–190.)* Individual II-2 has inherited the A allele from his mother (since she is homozygous for this allele) and the B allele from his father, the affected individual. In turn, he has passed on this B allele to the fetus, individual III-4. Since we do not know which grandpaternal allele is linked to the disease, and individual II-2 may be too young to be showing any signs or symptoms of the disease, we can only say that the fetus is at the same risk as his father, i.e., 50 percent.

377. The answer is B. *(Gelehrter, pp 202–207. Thompson, 5/e. pp 178–190.)* Individual II-2 has inherited the A allele from his unaffected mother and the B allele from his affected father. Individual III-1 has, in turn, inherited the A allele from her father. Assuming that no recombination has occurred, individual III-1 is unaffected.

378. The answer is A. *(Gelehrter, pp 202–207. Thompson, 5/e. pp 178–190.)* Individual II-2 has inherited the B allele from his father. Since he is now showing evidence of the disorder, it can be determined that this B allele is linked to the disease gene. The fetus, individual III-4, has inherited this B allele from his father. Therefore, assuming no recombination, the fetus must be affected.

379. The answer is B. *(Gelehrter, pp 202–207. Thompson, 5/e. pp 178–190.)* Cystic fibrosis is an autosomal recessive disorder that affects the lungs and the exocrine pancreas. The proband in this pedigree (individual II-2) has inherited the A allele from each parent. The fetus has inherited one A allele, which is linked to the disease, and one B allele, which is not linked to the disease. Therefore, the fetus is a carrier of cystic fibrosis but is not affected with the disease.

380. The answer is C. *(Gelehrter, pp 202–207. Thompson, 5/e. pp 178–190.)* Cystic fibrosis is the most common autosomal recessive disorder among whites. The proband in this pedigree (individual II-2) has inherited the A allele from each parent. Therefore, assuming no recombination, the A allele must be linked to the disease gene in both parents.

Individual II-1 has inherited the B allele from each parent and is therefore unaffected and a noncarrier of the disease gene.

381. The answer is D. *(Gelehrter, pp 202–207. Thompson, 5/e. pp 178–190.)* In this case, the affected individual (II-2) inherited the A allele from one parent and the B allele from the other. However, it cannot be determined which allele came from which parent. The fetus has also inherited the AB genotype; again, however, it cannot be determined which allele came from which parent. Therefore, the disease status of the fetus cannot be determined.

382. The answer is B. *(Gelehrter, pp 202–207. Thompson, 5/e. pp 178–190.)* The affected individual (II-2) has inherited the A allele from one parent and the B allele from the other. However, it is impossible to determine which allele came from which parent. Individual II-1 has inherited the AA genotype. Since she has inherited an A allele from both parents, it can be determined that one of those two alleles must be linked to the disease gene, although it cannot be determined from which parent it was inherited. The other A allele is not linked to the disease gene. The individual is therefore an unaffected carrier of this autosomal recessive condition.

383. The answer is D. *(Gelehrter, pp 69–121. Thompson, 5/e. pp 97–114.)* Allele-specific oligonucleotides (ASO) allow definitive genotyping by dot-blot analysis of patient DNA. For a two-allele system, identical amounts of patient DNA are dotted in duplicate on two membranes, then hybridized separately to individual ASO. After hybridization and washing to remove unbound ASO, the membranes are autoradiographed (for colorimetric probes, the appropriate developing reagents are used). After calibration with known genotypes, dot-blot analysis with ASO allows rapid genotype screening. If the genes are first amplified by polymerase chain reaction (PCR), dot-blot analysis can also be applied to blood spots obtained for newborn metabolic screening.

384. The answer is C. *(Gelehrter, pp 76–88, 102–106. Thompson, 5/e. pp 106–113.)* Sickle cell anemia is an autosomal recessive hemoglobinopathy that occurs in approximately 1 in 500 births in the black population. It is caused by a single nucleotide substitution in codon 6 of the hemoglobin gene. This mutation abolishes a restriction site for the enzyme MstII and thus can be detected by Southern blot analysis after digestion with this enzyme. It can similarly be detected when a PCR product is digested with this enzyme. ASO hybridization also detects single base pair substitutions. DNA sequencing will detect any change in the order of bases in a DNA fragment. Western blotting is a technique that examines the size and amount of mutant protein in cell extracts and therefore would not identify the sickle cell mutation in DNA.

385. The answer is B. *(Thompson, 5/e. p 92.)* Imprinting is a phenomenon in which the sex of the transmitting parent may affect the expression of the gene (the phenotype). These differences appear to be associated with differences in methylation patterns of DNA.

386. The answer is D. *(Gelehrter, p 15. Thompson, 5/e. pp 46–47.)* The "genetic code" uses three base "words," or codons, to specify the 20 different amino acids. Since there are four nucleotides that can be arrayed in 2^4 different combinations (64 total), the code must be degenerate, with more than one codon for a single amino acid. Sixty-one codons code for amino acids; three codons are "stop" codons and result in chain termination. The genetic code is universal with codons coding for the same amino acids in all organisms.

387. The answer is A. *(Gelehrter, p 44. Thompson, 5/e. pp 89–90.)* Mitochondrial inheritance is the transmission of genes encoded on the mitochondrial genome; Mendelian inheritance is the transmission of genes encoded on the nuclear genome. Mitochondria, located in the cytoplasm of cells, are maternally transmitted. Therefore, any disorder coded for on the mitochondrial genome is transmitted through the maternal line only. Affected males cannot pass on the disorder. Unlike X-linked recessive disorders, equal numbers of males and females are affected.

388. The answer is C. *(Gelehrter, p 44. Thompson, 5/e. pp 89–90.)* Mitochondrial DNA is a double-stranded, closed, circular molecule of 16,569 base pairs. It encodes its own tRNAs and encodes 13 proteins involved with oxidative phosphorylation, which are translated within the mitochondrion. The mitochondrial DNA is inherited from the mother.

389–392. The answers are: 389-D, 390-B, 391-E, 392-A. *(Thompson, 5/e. pp 97–114, 427–442.)* Nucleic acid hybridization involves the reassociation of DNA-DNA, DNA-RNA, or RNA-RNA complementary strands under controlled conditions. Inclusion of a radioactive or chemically labeled DNA or RNA segment (probe) allows quantitation of hybridization at the end of the reaction. The rate of hybridization partly reflects the complexity or total number of base pairs in the nucleic acid sample. Stringency of hybridization refers to the salt or temperature conditions employed; low-salt or high-temperature (high-stringency) conditions require perfect matching of the complementary strands before hybridization will occur. Oligonucleotides are short (15 to 50 base pairs), artificially synthesized DNA or RNA segments that may be employed as primers to initiate DNA sequencing or polymerase chain reactions (PCRs). The short length may also be used to advantage in stringent hybridization reactions to discriminate among alleles differing by a single base pair (allele-specific oligonucleotides).

393–396. The answers are: 393-C, 394-E, 395-A, 396-B. *(Gelehrter, pp 69–121.)* DNA polymorphisms (nucleotide sequence variations) occur approximately once per 200 to 500 base pairs (bp) of human DNA. If the sequence variation affects the recognition site for a restriction endonuclease, the altered segment sizes produced by endonuclease digestion allow detection of the sequence change (RFLP). Single-copy RFLPs become diagnostic of a chromosome region in a particular individual and are useful for family studies. Linkage (cosegregation) of the RFLP and a disease allele provides a measure of genetic distance between these loci: 1 percent recombinants = 1 cM = approximately 10^6 bp of DNA. Variable numbers of tandem repeats (VNTRs) often occur in repetitive DNA and yield numerous restriction fragment alleles. VNTRs, including their subfamilies of minisatellites and CA repeats, are thus very useful for DNA fingerprinting. Both RFLPs and VNTRs can now be simply detected by PCR (polymerase chain reaction using a thermal cycler) rather than Southern blotting.

397–401. The answers are: 397-C, 398-E, 399-A, 400-D, 401-B. *(Gelehrter, pp 69–121. Thompson, 5/e. pp 97–114.)* The 6×10^9 bp of DNA in each human diploid cell is apportioned among 46 chromosomes. Even with the highest resolution karyotype, the average chromosome band equals about 2×10^6 bp. These measurements emphasize the vastly greater precision of molecular analysis in detecting gene deletions (40,000 bp) or codon deletions as in the ΔF_{508} mutation (deletion of a phenylalanine codon) that is common in cystic fibrosis.

402–404. The answers are: 402-D, 403-E, 404-A. *(Thompson, 5/e. pp 97–114.)* With the advent of DNA cloning technology, it has become possible to saturate the genome with polymorphic DNA markers— anonymous segments of DNA that are cloned, tested for their ability to detect RFLPs, and mapped to specific chromosomal regions. These DNA probes are named according to their chromosome of origin, their order of discovery, and their ability to hybridize with single- or multiple-copy sequences. Once DNA marker studies have established linkage of a disease with a particular chromosomal band, nearby genes are identified based on their representation in libraries of complementary DNA (cDNA) clones.

405–409. The answers are: **405-B, 406-D, 407-C, 408-A, 409-B.** *(Gelehrter, pp 9–15. Thompson, 5/e. pp 32–33, 40–41.)* DNA (deoxyribonucleic acid), which encodes our genetic information, is a double helix that contains the four bases adenine, guanine, cytosine, and thymine. RNA (ribonucleic acid) differs from DNA in that it contains the sugar ribose instead of deoxyribose. In addition, uracil replaces the base thymine. mRNA or messenger RNA is the template upon which polypeptide chains are transcribed. Small (80-nucleotide) tRNAs (transfer RNAs) are specific for each amino acid sequence, which allows the appropriate amino acid to be added along the mRNA. Ribosomes contain rRNA, upon which mRNA translation into a polypeptide occurs.

410–413. The answers are: **410-B, 411-A, 412-C, 413-D.** *(Gelehrter, pp 9–17. Thompson, 5/e. pp 40–41.)* DNA is duplicated in a process called *replication*. The annealing of two complementary strands of DNA is called *hybridization*. Double-stranded DNA is transcribed into single-stranded mRNA (messenger RNA), which can then be translated with the help of amino acid–specific tRNAs (transfer RNAs) into polypeptide chains. Recombination is the formation of new linked genes via crossing-over.

414–418. The answers are: **414-A, 415-E, 416-B, 417-A, 418-D.** *(Gelehrter, pp 69–121. Thompson, 5/e. pp 97–114.)* Exons are the coding portions of genes and consist of trinucleotide codons that guide the placement of specific amino acids into protein. Introns are the noncoding portions of genes that may function in evolution to provide "shuffling" of exons to produce new proteins. The primary RNA transcript contains both exons and introns, but the latter are removed by RNA splicing. The 5′ (upstream) and 3′ (downstream) untranslated RNA regions remain in the mature RNA and are thought to regulate RNA transport or translation. A poly(A) tail is added to the primary transcript after transcription, which facilitates transport and processing from the nucleus. The discovery of introns complicated Mendel's idea of the gene as the smallest hereditary unit—a modern definition might be the colinear sequence of exons, introns, and adjacent regulatory sequences that accomplish protein expression.

419–423. The answers are: 419-A, 420-B, 421-D, 422-C, 423-D. *(Gelehrter, pp 193–277. Thompson, 5/e. pp 167–199.)* Analysis of segregation in families allows assignment of genes to particular chromosome regions and provides genetic distance in centimorgans (percentage recombination). Since microscopically equivalent chromosome regions can be variably condensed, and since recombination rates vary by chromosome region and sex, genetic distances are not precisely convertible to base pairs of DNA. Physical methods include direct visualization of genes on chromosomes by in situ hybridization, and calibration of intergene distance by localization of two genes to the same large restriction fragment (pulsed-field gel electrophoresis) or cloning vector (yeast artificial chromosomes, cosmids).

424–427. The answers are: 424-C, 425-B, 426-A, 427-D. *(Gelehrter, pp 76–88. Thompson, 5/e. pp 105–112.)* Southern blotting is a technique first described by Edward Southern. DNA fragments are separated on agarose gels by electrophoresis and then transferred to nitrocellulose filters. The filters are then exposed to labeled probes that hybridize to the DNA fragments. Northern blotting is an analogous procedure that allows for the detection of RNA fragments. The polymerase chain reaction (PCR) is a technique in which thermal cycling is used to rapidly amplify small segments of DNA.

428–431. The answers are: 428-C, 429-B, 430-C, 431-A. *(Gelehrter, pp 11–12. Thompson, 5/e. pp 31–33.)* DNA is composed of a backbone of deoxyribose sugars joined together by 5'-3' phosphodiester bonds. Attached to each sugar is a purine (adenine and guanine) or pyrimidine (thymine and cytosine) base. The unit of sugar, phosphate, and base constitutes a nucleotide. Two complementary strands of DNA are bound together by hydrogen bonds between adenine and thymine or between guanine and cytosine. In RNA, the sugar ribose replaces deoxyribose.

432–436. The answers are: 432-B, 433-C, 434-D, 435-D, 436-A. *(Gelehrter, pp 69–94. Thompson, 5/e. pp 40–51.)* In general, genes contain coding sequences, or exons, which are interrupted by intervening sequences, or introns. At the 5', or upstream, flanking region of the gene is a promoter region that regulates transcription. One of these regulatory elements is the CAT box (actually CCAAT). At the 3', or downstream, end of the gene is the poly(A) tail, which includes the sequence AATAAA and signals for the polyadenylation of mRNA, which increases its stability. Intronic sequences are removed during the processing of RNA. Pseudogenes are segments of DNA that closely resemble genes but are not functional.

437–441. The answers are: 437-D, 438-A, 439-B, 440-E, 441-E. *(Gelehrter, pp 17–18. Thompson, 5/e. pp 33–36.)* DNA is "packaged" via an efficient method of coiling. It is wound around proteins called *histones,* which are rich in basic amino acids. Together these condensed packages of DNA and protein are called *chromatin.* Heterochromatin is chromatin that contains repetitive DNA and stains darkly with trypsin and Giemsa. The centromere, also known as the *primary constriction,* is a region of heterochromatin by which sister chromatids are held together. Transposons are pieces of DNA that can move to a new insertion site in DNA.

442–446. The answers are: 442-E, 443-A, 444-B, 445-D, 446-C. *(Gelehrter, pp 69–121. Thompson, 5/e. pp 97–114.)* Insertion (question 444) or deletion (question 445) of nucleotides will shift the reading frame unless the change is a multiple of 3 (question 446). Frameshifts may create unintended stop codons as in question 444. Point mutations that result in nucleotide or amino acid substitutions are conveniently named by their position in the protein, e.g., $P_{21}A$, or DNA reading frame, e.g., $C_{63}A$. Deletions are prefixed by delta as with ΔF_{24}.

447–450. The answers are: 447-D, 448-B, 449-D, 450-C. *(Gelehrter, pp 193–277. Thompson, 5/e. pp 167–199.)* In order for the fetal blood type to be predictive of affliction with nail-patella syndrome (NPS), both the maternal and fetal genotypes must be informative. In other words, the family genotypes must unambiguously demonstrate which grandpaternal chromosome was transmitted to the fetus. Affliction of both father and daughter with NPS indicates that the chromosome carrying the O allele also has the abnormal NPS allele. For questions 447 through 449, the fetal blood type distinguishes which maternal chromosome was transmitted and is predictive of a 90 percent or 10 percent risk for NPS as opposed to the a priori 50 percent risk. These figures reflect the 10 percent probability of recombination. For question 450, it is unclear which maternal allele was transmitted to the fetus, so the a priori risk cannot be modified.

451–457. The answers are: 451-D, 452-B, 453-A, 454-C, 455-E, 456-A, 457-A. *(Gelehrter, pp 69–121. Thompson, 5/e. pp 97–114.)* The G-to-T mutation in r alleles ablates the E2 restriction site, which yields a distinctive 5-kb E1–E3 restriction fragment after Southern analysis. Probes that visualize the E1–E3 region will thus distinguish RR (3- and 2-kb fragments), Rr (5-, 3-, and 2-kb fragments), and rr (5-kb fragment only) genotypes. Probes that visualize restriction fragments outside of this region, such as the E3–E4 probe, will show no differences between genotypes.

458–462. The answers are: 458-D, 459-C, 460-B, 461-D, 462-B. *(Gelehrter, pp 69–121. Thompson, 5/e. pp 97–114.)* To determine fetal risk, maternal and fetal genotypes must be informative—first as to which RFLP allele cosegregates with the abnormal allele for NPS and second as to which maternal allele was inherited by the fetus. These conditions are met for the husband's and fetus' genotypes in questions 458, 460, and 461. Note that question 462 is a case of nonpaternity, but the absence of a 5 allele indicates that the fetus received mother's 3,2 allele and has a 95 percent risk for NPS.

Integrated Cases and Ethics

DIRECTIONS: Each question below contains four or five suggested responses. Select the **one best** response to each question.

Questions 463–473

A child is referred for evaluation of developmental delay and unusual behavior. On examination, you note loose joints, large ears, prominent jaw, and large testes for age.

463. This child has a

(A) sequence
(B) syndrome
(C) disruption
(D) single birth defect
(E) deformation

464. A karyotype should be considered for the child because

(A) the majority of children with mental retardation have chromosomal aberrations
(B) mental retardation (developmental delay) and multiple major or minor anomalies are hallmarks of chromosomal disease
(C) the patient has several characteristics of Down syndrome
(D) Klinefelter syndrome (47,XXY) is associated with megalotestes

465. The karyotype obtained from this child with developmental delay was first read as normal. Another physician, suspicious of a particular disease, ordered a second karyotype performed in special cell medium and obtained the result 46,XY,fra(X). The required cell medium was

(A) high calcium medium
(B) low calcium medium
(C) low phosphate medium
(D) low folic acid medium
(E) high folic acid medium

466. A family history was obtained for this child with developmental delay. The child's mother had learning disability and could only finish the ninth grade. Two maternal uncles had more severe mental retardation and required special education. The full pedigree is shown in the figure below, with filled symbols representing severe mental retardation and hatched symbols representing learning disability. The most likely inheritance pattern is

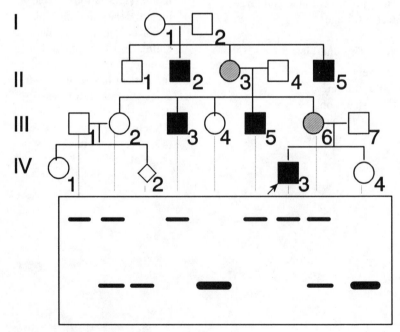

(A) autosomal recessive
(B) autosomal dominant
(C) X-linked recessive
(D) mitochondrial

467. The diagnosis is fragile X syndrome. Given the pedigree illustrated, individual III-6 has what risk of a severely affected male in a future pregnancy?

(A) 100 percent
(B) 50 percent
(C) 25 percent
(D) 12.5 percent
(E) Virtually 0

468. What is the risk that individual III-4 will have a severely affected male?

(A) 100 percent
(B) 50 percent
(C) 25 percent
(D) 12.5 percent
(E) Virtually 0

469. What is the risk that individual IV-3 will have a severely affected male?

(A) 100 percent
(B) 50 percent
(C) 25 percent
(D) 12.5 percent
(E) Virtually 0

470. In order to determine their risk for transmitting fragile X syndrome, karyotyping would be useful for which individuals in the illustrated pedigree?

(A) III-2, III-4, IV-1, IV-4
(B) III-2, III-4
(C) III-2, III-4, III-6,
(D) III-2, III-4, III-6, IV-1, IV-4
(E) IV-1, IV-3, IV-4

471. Because older female carriers of fragile X syndrome often do not exhibit the fragile site, alternative diagnostic testing is needed to determine their carrier status. The hypothetical Southern blot shown under the illustrated pedigree displays the DNA restriction fragments visualized with the probe DXS548 corresponding to individuals from the pedigree. DXS548 is a single-copy polymorphic locus that has been mapped to within 1 megabase of the fra(X) site. Based on these data, what is the risk that individual III-2 will have a son with the fragile X syndrome?

(A) 100 percent
(B) 50 percent
(C) 25 percent
(D) 12.5 percent
(E) Virtually 0

472. Individual III-2 of the pedigree and blot illustrated has requested prenatal diagnosis for her 16-week pregnancy. A fetal karyotype did not show the fragile X marker, and the DNA analysis is shown in the figure under individual IV-2. Based on these results, the most likely prenatal diagnosis and its accuracy are

(A) normal female, 95 percent
(B) carrier female, 95 percent
(C) unaffected male, 99 percent
(D) affected male, 99 percent

473. Suppose that individual
III-2 had become estranged from
her family and that her relatives
refused to give permission for re-
lease of their positive fragile X
testing results. Which of the following ethical principles and possible outcomes would the family
physician confront if he or she
notified individual III-2 of her
risks?

(A) Informed consent, medical
    commendation
(B) Informed consent, legal lia-
    bility
(C) Patient confidentiality, medi-
    cal commendation
(D) Patient confidentiality, legal
    liability
(E) None of the above

Questions 474–478

474. A child is referred for evaluation because of low muscle tone and
developmental delay. Shortly after delivery the child was a poor
feeder and had to be fed by tube. Later in the second year, the child
began to eat voraciously and became obese. He also had a slightly unusual face with almond-shaped eyes and down-turned corners of the
mouth. The hands, feet, and penis were small and the scrotum was
poorly formed. The diagnostic category and laboratory test to be considered for this child are

(A) sequence, serum testosterone
(B) single birth defect, serum testosterone
(C) deformation, karyotype
(D) syndrome, karyotype
(E) disruption, karyotype

475. A karyotype was performed on the obese child and was entirely normal. Because his physician suspected a disorder known as Prader-Willi syndrome, Southern blotting was performed to determine the origin of the patient's number 15 chromosomes. In part A of the figure below, a hypothetical Southern blot with DNA probe D15S8 defines which of four restriction fragment length polymorphisms (RFLPs) are present in DNA from mother (M), child (C), and father (F). Based on the D15S8 locus, what is the origin of the child's two number 15 chromosomes?

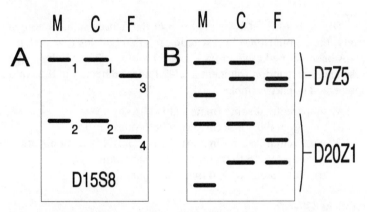

(A) One from the mother, one from the father
(B) Both from the father
(C) Both from the mother
(D) Results cannot be interpreted

476. Because part A of the figure demonstrates the child is missing both paternal chromosome 15 alleles, nonpaternity would be a more plausible explanation than uniparental disomy. The hypothetical Southern blot shown in part B illustrates a DNA "fingerprinting" analysis to examine paternity, where maternal (M), child (C), and paternal (F) DNA samples have been restricted, blotted, and hybridized simultaneously to the probes D7Z5 and D20Z1. The distribution of restriction fragment alleles suggests

(A) the child is adopted
(B) false maternity (i.e., baby switched in the nursery)
(C) false paternity
(D) correct maternity and paternity
(E) none of the above

477. As discussed in the previous question, the DNA analyses in the figure have the potential to demonstrate nonpaternity. If the physician ordering these analyses did not discuss this possibility with the mother, he or she would be in violation of

(A) patient confidentiality
(B) patient rights
(C) informed consent
(D) standards of care
(E) malpractice guidelines

478. Given that the analysis in part B of the figure excludes nonpaternity, the chromosome 15 restriction fragment alleles revealed in part A become predictive of particular chromosome anomalies. The presence of restriction fragment alleles 1 and 2 detected in the child as represented in part A implies

(A) uniparental disomy (maternal isodisomy 15)
(B) uniparental disomy (maternal heterodisomy 15)
(C) uniparental disomy (maternal isodisomy 15) or monosomy 15
(D) trisomy 15 due to maternal nondisjunction
(E) trisomy 15 due to paternal nondisjunction

Questions 479–482

479. The genesis of Prader-Willi syndrome by inheritance of two normal chromosomes from a single parent is an example of

(A) germinal mosaicism
(B) genomic imprinting
(C) chromosome deletion
(D) chromosome rearrangement
(E) anticipation

480. A child has severe epilepsy with fits of laughing, developmental delay, short stature, and autistic behavior. Her appetite is normal. The family history revealed three normal siblings, normal parents, no consanguinity, and two normal aunts and uncles on each side of the family. The family history rules out

(A) autosomal recessive disease
(B) autosomal dominant disease
(C) X-linked dominant disease
(D) chromosomal disease
(E) multigenerational disease

481. The child in the previous question has characteristic manifestations of a condition known as the *Angelman syndrome*. Because of the syndromic nature of the disorder and the developmental delay, a karyotype is performed and yields the chromosomes 15 shown below. The partial karyotype reveals

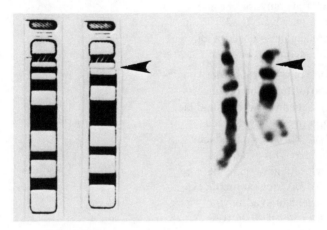

(A) interstitial deletion of 15q
(B) interstitial deletion of 15p
(C) pericentric inversion of 15
(D) paracentric inversion of 15
(E) none of the above

482. A Southern analysis is performed on the patient discussed in the previous two questions using the probe D15S8. Assume that the parental restriction fragment alleles were the same as those in the figure at question 475. What conclusion could be drawn if the child's analysis revealed the presence of only allele 3?

(A) Paternal heterodisomy 15
(B) Biparental contribution of chromosomes 15, D15S8 locus within deleted region, deletion occurred on paternal 15
(C) Biparental contribution of chromosomes 15, D15S8 locus within deleted region, deletion occurred on maternal 15
(D) Biparental contribution of chromosomes 15, D15S8 locus not within deleted region
(E) None of above

483. Questions 474 through 482
illustrated that a patient with
deletion of the maternal 15q13q15
region may have Angelman syndrome and a patient with maternal heterodisomy for chromosome 15 may have Prader-Willi
syndrome. These observations
may be reconciled because

(A) lack of a paternally derived
     15q13q15 region produces
     Angelman syndrome and lack
     of a maternally derived
     15q13q15 region produces
     Prader-Willi syndrome
(B) lack of a maternally derived
     15q13q15 region produces
     Angelman syndrome and lack
     of a paternally derived
     15q13q15 region produces
     Prader-Willi syndrome
(C) deletion of 15q13q15 pro-
     duces Angelman syndrome
     while uniparental disomy 15
     produces Prader-Willi syn-
     drome
(D) deletion of 15q13q15 pro-
     duces Prader-Willi syndrome
     while uniparental disomy 15
     produces Angelman syn-
     drome

Questions 484–490

An infant presents at day 10 of life with poor feeding, vomiting, and lethargy.

484. The differential diagnosis would include all the following EXCEPT

(A) syndrome
(b) sepsis
(C) cardiomyopathy
(D) CNS catastrophy
(E) inborn error of metabolism

485. Further history reveals that the baby was the full-term product of a normal pregnancy and delivery. He did well in the first 48 h of life and was discharged home with his mother. The baby then developed progressive lethargy, anorexia, and vomiting over the next week. In order to evaluate the possibility of an inborn error of metabolism, you ask the parents about all the following EXCEPT

(A) family history of neonatal deaths
(B) family history of consanguinity
(C) infant's feeding history
(D) any unusual odors
(E) their karyotypes

486. Laboratory tests reveal a low WBC count, metabolic acidosis, increased anion gap, and mild hyperammonemia. Measurement of plasma amino acids reveals elevated levels of glycine and measurement of urinary organic acids reveals increased amounts of propionic acid and methyl citrate. A diagnosis of propionic acidemia is made. Family history reveals that the parents are first cousins. This disorder is likely to be

(A) autosomal dominant
(B) autosomal recessive
(C) X-linked dominant
(D) X-linked recessive
(E) multifactorial

487. Biotin is a cofactor in the reaction catalyzed by PCC as well as several other carboxylases. In holocarboxylase synthase deficiency, none of these carboxylases can be made. A similar picture may be seen in biotinidase deficiency, a disorder of biotin metabolism. These disorders constitute an example of

(A) allelic heterogeneity
(B) genetic heterogeneity
(C) variable expressivity
(D) incomplete penetrance
(E) dosage compensation

488. Propionic acidemia is caused by a deficiency of propionyl-CoA carboxylase (PCC) as shown in the following reaction:

Valine
Odd-chain fatty acids   PCC
Methionine   →   propionyl-   ─↛   methylmalonyl-CoA
Isoleucine   CoA
Threonine   ↓
                         propionic acid
                         methyl citrate

In the treatment of this disorder, contraindications would include?

    (A) antibiotics
    (B) a diet high in valine, isoleucine, and methionine
    (C) caloric supplementation
    (D) aggressive fluid and electrolyte management
    (E) hemodialysis

489. The parents of the proband in this family return to your office 2 years later. The mother of the proband is pregnant. The risk that the fetus is affected with propionic acidemia is

(A) 2/3
(B) 1/2
(C) 1/4
(D) 1/8
(E) virtually 0

490. DNA analysis is performed on the family. Results are shown in the figure below. The risk that the fetus is affected with propionic acidemia is now

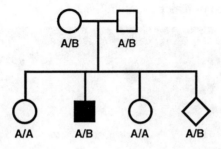

(A) virtually 100 percent
(B) 2/3
(C) 1/2
(D) 1/4
(E) virtually 0

Questions 491–494

A newborn presents with ambiguous genitalia as seen in the figure below. The photograph shows an enlarged clitoris or small phallus with labial fusion or hypoplastic scrotum.

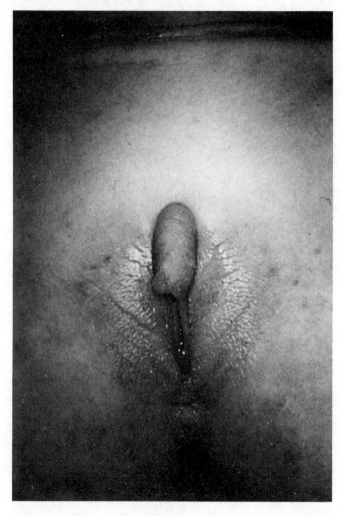

491. This child's sex is most reliably established by

(A) buccal smear to determine if there are one or two Barr bodies
(B) buccal smear to determine if there is one Barr body or none
(C) peripheral blood karyotype
(D) bone marrow karyotype
(E) polymerase chain reaction (PCR) using primers specific for the Y chromosome long arm

492. The dot-blot shown below examines the proband's DNA after PCR amplification and hybridization with DNA probes from the X and Y chromosome. In this case, the Y chromosome probe is from the SRY region of Yp that has recently been characterized as the male-determining region. DNA from control male and female patients is also applied to the dot-blot. Based on the dot-blot results, which of the following conclusions can be reached?

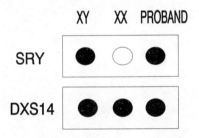

(A) The proband is a genetic male
(B) The proband is a genetic female
(C) The proband is male
(D) The proband is female
(E) None of the above

493. Based on the photograph and dot-blot analysis, the anomaly illustrated is

(A) female pseudohermaphroditism
(B) male pseudohermaphroditism
(C) true hermaphroditism
(D) an XY female
(E) an XX male

494. The physician received the karyotype result on the proband and it was 46,XY. The most likely diagnosis was local failure of the external genitalia to respond to testosterone, and this was supported by studies excluding defects of pituitary/adrenal function and of the internal genitalia. The physician chose not to tell the parents that their child was actually a genetic male, but emphasized that she was a female who should return for management at puberty because of the risk of sexual and reproductive problems. From the ethical perspective, this action would fall under the category of

    (A) patient confidentiality
    (B) nondisclosure
    (C) informed consent
    (D) failure to provide ongoing care
    (E) discrimination

DIRECTIONS: The group of questions below consists of lettered headings followed by a set of numbered items. For each numbered item select the **one** lettered heading with which it is **most** closely associated. Each lettered heading may be used **once, more than once, or not at all.**

Questions 495–500

Match the chromosomal anomalies below with their phenotypes.

(A) Normal external female genitalia, ovarian dysgenesis, webbed neck, short stature
(B) Possible presentation as the proband in the figure accompanying questions 491 through 494
(C) Tall stature, eunuchoid habitus, small testes
(D) Normal female external appearance
(E) Normal male external appearance

495. Turner syndrome

496. Klinefelter syndrome

497. Chimerism (46,XX/46,XY)

498. 47,XXX/46,XX mosaicism

499. 46,X,i(Xq)

500. 47,XYY

Integrated Cases
and Ethics
Answers

463. The answer is B. *(Thompson, 5/e. p 391.)* A pattern of major or minor anomalies is known as a *syndrome*. The child's minor abnormalities plus developmental delay are strongly suggestive of a syndrome. It is important to distinguish syndromes from sequences (multiple consequences produced by a single embryologic error) or isolated birth defects (disruptions, deformations, malformations) since the latter categories usually have an optimistic prognosis with minimum recurrence risk.

464. The answer is B. *(Gelehrter, pp 171–176.)* Mental retardation of unknown cause, particularly when accompanied by multiple major or minor anomalies, is an indication for karyotyping. The child's facial appearance and minor anomalies are atypical of Down syndrome. Males with Klinefelter syndrome have small rather than large testes.

465. The answer is D. *(Gelehrter, pp 185–187. Thompson, 5/e. pp 81–82.)* The fra(X) notation indicates a discontinuity near the tip of the X long arm. Cytogenetic visualization of this fragile site requires lymphocyte culture in low folic acid medium before metaphase arrest. Originally noticed by chance, the 46,XY,fra(X) karyotype is now associated with a fragile X syndrome of mental retardation, loose joints, characteristic face, and megalotestes.

466. The answer is C. *(Gelehrter, pp 39–45. Thompson, 5/e. pp 75–80.)* The pedigree demonstrates oblique transmission with severely affected males and normal or mildly affected females. X-linked recessive inheritance is most likely, although X-linked dominant inheritance with variable, milder expression in females is also possible. The lack of phenotypic findings in individual I-1 argues against X-linked dominant inheritance since the fact she has two affected sons indicates she has the abnormal allele.

467. The answer is C. *(Gelehrter, pp 39–45. Thompson, 5/e. pp 75–80.)* Individual III-6 is an obligate carrier because of her affected son. She has a 1/2 chance for a son and a 1/2 chance for the son to have fragile X syndrome, which results in a (1/2)(1/2) = 1/4 overall risk.

468. The answer is D. *(Gelehrter, pp 39–45. Thompson, 5/e. pp 75–80.)* Individual III-4 has a 1/2 chance to be a carrier and a 1/4 chance to have an affected son if she is a carrier. This results in a (1/2)(1/4) = 1/8 overall risk.

469. The answer is E. *(Gelehrter, pp 39–45. Thompson, 5/e. pp 75–80.)* Affected males, such as individual IV-3, cannot transmit their abnormal X chromosome allele to sons.

470. The answer is A. *(Gelehrter, pp 185–187. Thompson, 5/e. pp 81–82.)* Females may transmit abnormal X chromosome alleles but be unaffected themselves because of the normal X chromosome allele. Females at risk to be carriers of X-linked recessive diseases thus need counseling and detection of their carrier status where possible. Some female carriers of the fragile X chromosome will show the fragile X site by karyotyping under appropriate culture conditions.

471. The answer is C. *(Gelehrter, pp 193–207. Thompson, 5/e. pp 420–423.)* The restriction fragment allele of lower molecular weight (nearer the bottom of the hypothetical autoradiogram in the figure) is linked to the normal X chromosome allele. Individual III-2 is thus definitely a carrier. Phase (which RFLP allele segregates with the disease allele) is easily established in X-linked disorders by noting the restriction fragment present in affected males.

472. The answer is C. *(Gelehrter, pp 193–207. Thompson, 5/e. pp 420–423.)* The fetus has received the lower molecular weight restriction fragment allele from his mother and is at minimal risk to have fragile X syndrome. The fetus must be male since he has received neither paternal X chromosome allele. Accuracy of the diagnosis depends on the chance of recombination between the restriction fragment and fragile X loci. The distance between these loci is indicated as <1 megabase in the previous question. This distance is approximately equivalent to 1 centimorgan or a 1 percent chance of recombination.

473. The answer is D. *(Thompson, 5/e. p 395.)* Genetic counseling, like other medical practice, entails privileged communication between patient and physician. All information regarding the patient is confidential and cannot be shared with any institution or individual without the patient's written permission. Violation of this confidentiality, even for the most humane of reasons, places the physician at legal risk. Informed consent refers to a patient's knowledgeable agreement to undergo a medical procedure.

474. The answer is D. *(Thompson, 5/e. p 391.)* This child has several minor anomalies, a major anomaly that affects the genitalia, and developmental delay. These multiply affected and embryologically unrelated body regions suggest a syndrome rather than a sequence. Because of the multiple anomalies and developmental delay, the first diagnostic test to be considered would be a karyotype rather than a test for specific organ function such as serum testosterone.

475. The answer is C. *(Thompson, 5/e. pp 92–94.)* The hypothetical probe D15S8 would imply a unique DNA segment that recognizes a single locus on chromosome 15—the eighth such anonymous DNA probe to be isolated. Since normal individuals will have two number 15 chromosomes, they should have two alleles visualized after DNA restriction and hybridization with probe D15S8. Since both parents are heterozygous for the D15S8 locus as shown in part A of the figure, the child's result suggests that he has only received the maternal alleles (alleles 1,2) for locus D15S8. This implies that he has received both number 15 chromosomes from his mother. This is known as *uniparental disomy* and may occur by correction of trisomy 15 conceptions through loss of the paternal number 15 chromosome.

476. The answer is D. *(Thompson, 5/e. pp 129–132.)* DNA fingerprinting is used in both paternity and forensic analysis and relies on highly variable DNA polymorphisms called VNTRs (variable numbers of tandem repeats). The multicopy repeats include $(CA)_n$ and minisatellite sequences that are present throughout the genome. The usual VNTR probe is directed against single-copy DNA that flanks these repeats and yields multiple restriction fragment sizes that reflect the number of intervening repeats. The hypothetical probes D7Z5 and D20Z1 shown in part B of the figure recognize VNTR loci on chromosomes 7 and 20 that yield at least three alleles. Since the child's two alleles for D7Z5 (and D20Z1) match those of mother and father, correct maternity and paternity is established with a degree of error equal to the chance these allele combinations would occur in an unrelated individual. In practice, at least five VNTR probes are employed so that the odds for paternity (or nonpaternity) are very high indeed.

477. The answer is C. *(Thompson, 5/e. p 395.)* Informed consent requires that the patient be informed of all adverse effects that might result from a procedure. Evidence for nonpaternity may result from various types of DNA analysis and should be discussed with the concerned parties at the time of blood collection. Some physicians would speak to mother and father separately about this issue in order to maximize the opportunity for independent decision.

478. The answer is B. *(Thompson, 5/e. pp 92–94.)* Since the parents are heterozygous for differently sized D15S8 restriction fragment alleles, each allele becomes a marker for the respective parental 15 chromosome. The presence of alleles 1 and 2 in the child implies uniparental disomy with both maternal 15 chromosomes being passed to the child. This would represent heterodisomy, with both of the mother's number 15 chromosomes being passed down, as opposed to isodisomy where two copies of one maternal chromosome are transmitted. Uniparental isodisomy would make the child homozygous for a single restriction fragment allele and could not be distinguished from monosomy unless dosage or cytogenetic studies were performed. The presence of three chromosome 15 alleles would imply trisomy for the number 15 chromosome, and the particular parental alleles inherited would reveal the origin of the nondisjunction.

479. The answer is B. *(Thompson, 5/e. pp 92–94.)* In humans and other mammals, the source of genetic material may be as important as its content. Mice manipulated to receive two male pronuclei develop as abortive placentas, while those receiving two female pronuclei develop as abortive fetuses. The different impact of the same genetic material according to whether it is transmitted from mother or father is due to genomic imprinting. The term *imprinting* is borrowed from animal behavior and refers to parental marking during gametogenesis—the physical basis may be DNA methylation or chromatin phasing. Both maternally derived and paternally derived haploid chromosome sets are thus necessary for normal fetal development—this is why parthenogenesis does not occur in mammals. The imprint is erased in the fetal gonads and reestablished based on fetal sex. Certain cases of Prader-Willi syndrome are disorders of imprinting with absence of the paternally imprinted chromosome 15.

480. The answer is E. *(Gelehrter, pp 255–262. Thompson, 5/e. pp 53–94.)* The isolated or sporadic case is a problem in genetic counseling because the pedigree is not helpful in determining recurrence risk. New mutations for autosomal or X-linked dominant disorders, new chromosomal rearrangements, autosomal recessive conditions, and nongenetic disorders may all present as the first case in a family. A negative family history does rule out multigenerational disease.

481. The answer is A. *(Gelehrter, pp 165–171. Thompson, 5/e. pp 207–208.)* The arrow in the figure points to the deletion of region 15q11q13 in one of the patient's number 15 chromosomes. This interstitial deletion is seen in approximately 50 percent of patients with Prader-Willi and Angelman syndromes. It is now known that imprinting of the 15q11q13 region is important in the genesis of these two syndromes.

482. The answer is C. *(Thompson, 5/e. pp 92–94.)* The presence of two chromosomes 15 in the partial karyotype excludes monosomy 15 as an answer. Paternal isodisomy is a possible answer, but paternal heterodisomy would require the presence of alleles 3 and 4. Presence of the D15S8 locus within the 15q13q15 deleted region is most likely based on the cytogenetic result; this implies that the maternal chromosome 15 was deleted since neither maternal allele is transmitted to the child.

483. The answer is B. *(Thompson, 5/e. pp 92–94.)* Since approximately 50 percent of patients with Prader-Willi or Angelman syndrome have identical deletions of 15q13q15, deletion alone cannot explain the differences between these syndromes. DNA probe analysis similar to that shown in the figure at question 475 has shown that the absence of 15q13q15 is on the maternally derived chromosome in Angelman syndrome and on the paternally derived chromosome in Prader-Willi syndrome. Uniparental disomy 15 from the mother would be equivalent to deleting the paternally imprinted 15q13q15 region and is found in Prader-Willi syndrome. Paternal uniparental disomy 15, though rare, has been demonstrated in Angelman syndrome.

484. The answer is A. *(Nelson, 2/e. pp 164–169.)* In the neonate, nonspecific symptoms such as poor feeding, lethargy, vomiting, respiratory distress, coma, and seizures may be signs of a multitude of disorders including sepsis; diseases of the cardiopulmonary, gastrointestinal, and central nervous systems; and inborn errors of metabolism. A syndrome is a pattern of major and minor malformations and is not described in this infant.

485. The answer is E. *(Nelson, 2/e. pp 164–169.)* Since most inborn errors of metabolism are recessive disorders, parental consanguinity or a history of neonatal deaths within the same sibship are important clues. In addition, because a few inborn errors of metabolism are X-linked, it is also important to ask about neonatal deaths on the mother's side of the family. Since inborn errors of metabolism may involve an inability to metabolize various components of food such as protein or fats, dietary history is extremely important. Several inborn errors also produce an unusual odor of the urine or sweat. Because inborn errors are single gene defects, a karyotype is not usually helpful in making the diagnosis.

486. The answer is B. *(Scriver, 6/e. pp 2083–2103.)* Most inborn errors of metabolism are inherited as autosomal recessive disorders. A family history of consanguinity would also suggest an autosomal recessive mode of inheritance.

487. The answer is B. *(Gelehrter, pp 32–36.)* In this example, mutations in different genes and at different loci can produce a similar phenotype. This phenomenon is known as *genetic heterogeneity*. In allelic heterogeneity, different mutations at the same locus may produce abnormal phenotypes. Penetrance is the all-or-none expression of the gene, while expressivity refers to the variation in degree of severity of the phenotype. Dosage compensation is a phenomenon related to X-inactivation.

488. The answer is B. *(Nelson, 2/e. pp 164–168.)* In treating inborn errors of metabolism that present acutely in the newborn period, aggressive fluid and electrolyte therapy and caloric supplementation are important in order to correct the imbalances caused by the disorder. Since many of the metabolites that build up in inborn errors of metabolism are toxic to the CNS, hemodialysis is recommended for any patient in stage II coma (poor muscle tone, few spontaneous movements, responsive to painful stimuli) or worse. Hemodialysis is 10 times as effective as peritoneal dialysis in removing toxic metabolites. Dietary therapy should minimize substances that cannot be metabolized—in this case valine, methionine, and isoleucine. Antibiotics are frequently useful since metabolically compromised children are more susceptible to infection.

489. The answer is C. *(Gelehrter, pp 202–207. Thompson, 5/e. pp 178–190.)* The recurrence risk for an autosomal recessive disorder is 1 in 4, or 25 percent.

490. The answer is C. *(Gelehrter, pp 202–207. Thompson, 5/e. pp 178–190.)* The proband in this case has inherited the A allele from one parent and the B allele from the other. However, it is impossible to determine which allele came from which parent. The fetus has the same genotype as his affected brother. However, it cannot be determined if he inherited these alleles from the same parents as the affected boy and is thus, himself, affected or from the opposite parents and thus is an unaffected noncarrier. It can be said that he is definitely not an unaffected carrier. Assuming no recombination has occurred, the risk for the fetus to be affected is 1/2, or 50 percent.

491. The answer is C. *(Gelehrter, pp 179–189. Thompson, 5/e. pp 243–245.)* A peripheral blood karyotype provides the most reliable examination of the sex chromosomes. A bone marrow karyotype is more rapid (it uses rapidly dividing bone marrow cells) but usually has less resolution for defining subtle X and Y chromosome rearrangements. A buccal smear would theoretically show one Barr body in females (representing inactivation of one X chromosome) and none in males. In practice, this test is not very reliable and is rarely used. Detection of material of the Y long arm by PCR would be useful but does not examine the Y short arm that contains the sex-determining region.

492. The answer is A. *(Gelehrter, pp 82–85, 171–176. Thompson, 5/e. pp 108–109, 243–245.)* The dot-blot demonstrates hybridization of the proband's DNA with the DXS14 and SRY DNA probes and establishes the diagnosis of a genetic male. Gender assignment is not based solely on genetic testing but must include surgical and reproductive prognoses for male versus female adult function. For these reasons, the patient with ambiguous genitalia is a medical emergency that requires delicate management until gender assignment is agreed upon. The proband, for example, was judged not to have adequate phallic tissue for reconstruction of normal male genitalia and underwent appropriate surgery for female gender assignment.

493. The answer is B. *(Gelehrter, pp 179–189. Thompson, 5/e. pp 243–245.)* True hermaphroditism implies the presence of both male and female genitalia in the same patient and is extremely rare. Male pseudohermaphroditism implies a genetic male with incomplete development of his genitalia, as in the proband. Causes can range from abnormalities of the pituitary-adrenal-gonadal hormone axis to local defects in tissue responsiveness to testosterone. The XY female and XX male refer to phenotypically normal individuals whose genetic sex does not match their phenotypic sex. Examples include testicular feminization and pure gonadal dysgenesis (XY females) and offspring of fathers with Y translocations that inherit a cryptic SRY region without a visible Y chromosome (XX males).

494. The answer is B. *(Gelehrter, pp 171–176. Thompson, 5/e. p 391.)* Nondisclosure is generally inappropriate for a physician-patient relationship unless certain facts are judged to violate the higher doctrine of causing the patient no harm. Nondisclosure is more frequently used as a manner of presentation than as a withholding of facts—e.g., by using terms such as "developmental delay" or "seriously ill" rather than emotionally charged terms such as "mental retardation" or "dying." In this case, the patient's necessary management as a female was considered most relevant for the parents so that her gender identity rather than genetic sex was stressed. Continuity of care is essential under such circumstances so that future medical problems can be appropriately managed.

495–500. The answers are: 495-A, 496-C, 497-B, 498-D, 499-A, 500-E. *(Thompson, 5/e. pp 243–245.)* Turner syndrome is associated with deficiency of an X chromosome in females with karyotypes such as 45,X, 45,X,i(Xq), or 45,X/46,XX mosaicism. Turner syndrome females have normal external genitalia with ovarian dysgenesis, short stature, and variable physical stigmata such as a webbed neck. Klinefelter syndrome (47,XXY) is associated with tall stature and small testes that become more obvious at the time of puberty. 47,XXX females and 47,XYY males are normal in their appearance prior to puberty; these karyotypes are associated with reproductive and possibly behavioral problems. Chimerism refers to rare instances in humans in which single fetuses develop from cells of two or more zygotes. If the zygotes are of different sex, 46,XX/46,XY chimerism results and can produce a variety of genital malformations including those exhibited by the proband. Somatic mosaicism such as 47,XXX/46,XX represents two chromosomally distinct cell lines derived from the same zygote.

Bibliography

Gelehrter TD, Collins FS: *Principles of Medical Genetics*. Baltimore, Williams & Wilkins, 1990.

Nelson NM: *Current Therapy in Neonatal-Perinatal Medicine,* 2/e. Philadelphia, BC Decker, 1989.

Scriver CR, Beaudet AL, Sly WS, Valle D: *The Metabolic Basis of Inherited Disease,* 6/e. New York, McGraw-Hill, 1989.

Thompson MW, McInnes RR, Willard HF: *Genetics in Medicine,* 5/e. Philadelphia, WB Saunders, 1991.

Notes

Notes

Notes

Notes

Notes

Notes

Notes